Pass the QTS Numeracy Skills Test with Ease

2018 - 2020

By Vali Nasser

E-book editions are also available for this title. For more information email: valinasser@gmail.com

ISBN-13: 978-1979924900

ISBN-10: 1979924902

New Edition: Nov 2017

About the Author

The author of this book has experience in both consultancy work and teaching. He is intimately familiar with the QTS Numeracy testing as he was project manager at OCR working in conjunction with the Teaching Agency in implementing the initial phase of the Numeracy Skills Testing project. As a specialist mathematics teacher, he has subsequently tutored and taught mathematics in schools as well as in adult education. Finally, the author's initial book 'Speed Mathematics Using the Vedic System' has a significant following and has been translated into Japanese and Chinese as well as German.

The author hopes that his new book 'Pass the QTS Numeracy Skills Test with Ease' for 2018 - 2020 will be beneficial to those who want to re-visit their maths for the purposes of passing the QTS Numeracy skills test.

Introduction

For entry into Initial Teacher Training courses in 2018 - 2020 the contents of QTS Numeracy Skills Test have not changed from the previous years. This means the content is the same as my previous book for 2016 – 2017 with just a few typos corrected. However, as you probably know since 2012 - 2013 **the pass mark for the QTS Numeracy test has been raised to 63 percent in total.** To be safe you should aim at getting a **minimum of 18 out of the 28 questions right in the actual test itself.** This book is aimed at helping you to pass the QTS Numeracy Skills Test the first time you take it. If you feel that you need to improve your speed in the Mental Arithmetic part of the test and re-visit some areas in general arithmetic as well as statistics then this book will prove very useful to you. It will be particularly helpful if you do not feel so confident in maths, got a Grade C or did your exams a long time ago.

Although a lot of the material in the first two chapters will be familiar to you, hopefully you will find some of the 'Speed Methods' introduced helpful particularly for the Mental Arithmetic part of the Numeracy test. In the Mental Arithmetic test, you will be given 12 questions, each question is read out two times, after the second reading you will have 18 seconds to answer the question. However, paper and pencil will be provided to do some working out and you can of course be thinking about working out the question after the first reading. It would be advisable to jot down your calculation step by step and work out the answers before typing in the answers. As you probably know, the mental arithmetic test will be audio based and cover areas of basic arithmetic including calculations using time, fractions, percentages, measurements and conversions Unfortunately, if you cannot answer a question you have to move on to the next one once the time is up. Also, you can't go back to any question you have missed. So working with speed as well as accuracy will be important.

In addition, you will have 16 On screen questions for which an On-screen calculator is provided. You will have 36 minutes for the On-screen tests which works out at just over 2 minutes per question. The contents of these tests include being able to identify trends correctly, interpret statistical information accurately and make comparisons in order to draw conclusions. In addition, you will be set problems which involve time, money, proportion and ratio, percentages, fractions and decimals, measurements, conversions and averages, including mean, median, mode and range where relevant. Finally, you are also expected to know how to use simple formulae. In the On-screen tests you can go backwards and forwards and answer questions in any order within the overall time limit specified.

The total time for both the Mental Arithmetic and the On-screen test is 48 minutes. There is good material available from teacher training educational websites including On-line practice tests whose website address is given later.

Hopefully, this book will be an additional resource for those wanting to make sure you can revise the basics and practice more tests. In total, this book has over a hundred questions together with detailed answers. This includes many examples as you work through the various topics as well as five Mock Mental Arithmetic Tests and two On-screen tests.

One thing to remember is there is often more than one way of working out a given problem. It does not matter which method you use, so long as you feel comfortable with it. You will be marked only for the correct answer. The arithmetic part of the book gives you a variety of methods to choose from, including 'Speed Methods' of calculations.

Finally, although some of you may find the first two chapters very easy and be tempted to skip them, my advice is, just go through them quickly to make sure you remind yourself of speed methods of adding and subtracting as well as those of multiplying and dividing without a calculator. Remember in the mental arithmetic part of the test you will be working against limited time. Good luck with your tests.

Chapter 1 Mental Arithmetic part I

Addition and Subtraction using Speed Methods

The normal approach of column addition and subtraction is a good method and if you feel happy with it then you should have no problems with this part of arithmetic. Make sure that when dealing with adding and subtracting decimal numbers, the decimal points are aligned.

The following additional methods will prove useful in increasing your speed with the mental arithmetic part of the QTS Numeracy test.

Consider the ***Speed Method*** below for addition

<u>Compensating or adjusting method</u>

In this method we simply adjust by adding or subtracting from the rounded up or rounded down number as shown in the examples below. In example1 we round up 96 to 100 and adjust by taking away 4. Similarly we round up 69 to 70 and adjust by taking away 1. See below for all the working out.

Ex1

96 + 69 =

100 – 4 + 70 – 1 =

170 – 5 = 165

Ex2

59 + 88 +23 =

60 – 1 + 90 – 2 + 20 + 3 =

150 + 20 – 3 + 3 = 170

Mental Arithmetic Question

A teacher buys three resource packs A, B and C. The costs are as follows: A costs £3.90, B costs £13.85 and C costs £11.95. Find the total cost of the three resource packs

Method:

Total cost = **£3.90 + £13.85 +£11.95**

= £4 - 10p +£14 - 15p + £12 – 5p = £4 +£14 +£12 - 10p - 15p - 5p

= £30 – 30p = £29.70

Subtraction

You probably remember column subtraction and the number line method from your GCSE days or from when you last did maths. Before we go on to use the ***'Speed Method'*** let us revisit the familiar method for subtraction.

Example1:

Work out: 241 - 28

Traditional column method

The traditional methods of subtraction serve us well in mathematics. However, there is one more strategy that we can use to make this process much easier but more of this later. First, we will consider the normal approach.

Consider the following example:

$$\begin{array}{r} 241 \\ -\ \ 28 \\ \hline 213 \\ \hline \end{array}$$

Starting from the right-hand side we cannot subtract 8 from 1 so we borrow 1 from the tens column to make the units column 11. Subtracting 8 from 11 gives us 3. However, since we have taken away 1 from the tens column we are left with 3 in this column. Subtracting 2 from 3 in the tens column gives us 1. Since we have nothing else to take away the final answer is 213.

Speed Method of Subtraction

Example1: Now consider the same problem using a *Speed Method.*

If we add 2 to the top and bottom number we get:

```
   243    (241+2)
 _  30    (28+2)
 ______
   213
```

You can see that subtracting 30 from 243 is easier than subtracting 28 from 241!

This strategy relies on the algebraic fact that if you add or subtract the same number from the top and bottom numbers you do not change the answer to the subtraction sum.

So essentially, we try and add or subtract a certain number to both the numbers in order to make the sum simpler. A few more examples will help.

Example 2:

```
 113
 - 6
 ___
```

Add 4 to both numbers (we want to try to make the units column 0 in the bottom row if we can and if it helps) So the new sum is:

```
 117
 - 10
 ____
 107
 ____
```

We can see that if we subtract 10 from 117 we get 107.

Example 3:

```
  321
- 114
```

Let us add 6 to each number so that the unit column in the bottom number becomes a 0 as shown below:

```
  327 (add 6 to 321)
- 120 (add 6 to 114)
  207
```

Subtracting 120 from 327 we get 207 as shown. No borrowing is required.

Note: Sometimes you might find the method above useful; at other times it is easier to revert to the traditional method.

Subtracting from 100, 1000, 10000, 100000

Some people find subtracting from 1000, 10000 or 100000 difficult, so let us consider a useful technique for doing this.

Subtracting from 100, 1000 or 10000 using a '*Speed Method*'

In this case we use the rule **'all from nine and the last from 10'**

Example 1: 100 -76

We simply take each figure (except the last) in 76 from 9 and the last from 10 as shown below:

```
  1 0 0
-   76
   2 4
```

Take 7 from 9 to give 2 and take 6 from 10 to give 4

Example 2: 1000 – 897 =103

We simply take each figure (except the last) in 897 from 9 and the last from 10 as shown below:

```
 1 0 0 0
 - 8 9 7
   1 0 3
```

(Take 8 from 9 to give 1. Take 9 from 9 to give 0 and take 7 from 10 to give 3)

Subtracting from 2000, 3000, 4000, 5000, or more thousands

From the above, use the principle of 'last from 10 and the rest from nine' and 'subtracting 1 from the first digit on the left after all the zeros'

Example 1: Work out 3000 – 347

Using the principle of 'last from 10 the rest from nine' and 'subtracting 1 from the first digit on the left after all the zeros'.

We get the answer to be 2653

Example 2: Work out 7000 – 462

Similarly, the answer in this case is 6538.

Mental Arithmetic Question

A science teacher has 10000 Milliliters of a particular liquid, she uses up 8743 Milliliters after several class experiments. How much does she have left?

Method:

$$\begin{array}{r} 10000 \\ -\ 8743 \\ \hline 1257 \\ \hline \end{array}$$

(Take 8 from 9 to give 1, 7 from 9 to give 2, 4 from 9 to give 5 and finally 3 from 10 to give 7)

This means the teacher has 1257 milliliters of liquid left.

Multiplying & Dividing by 10, 100 and 1000 (by powers of 10)

You are expected to be familiar with multiplying and dividing numbers by 10, 100, 1000 or any other power of 10

Speed Method: Rule for multiplying whole numbers:

(1) When multiplying a whole number by 10 add a zero at the end of the number.

(2) When multiplying by 100 add two zeros.

(3) When multiplying by 1000 add three zeros

(4) You simply add the number of zeros reflected in the power of 10.

Some examples will illustrate this:

(1) $45 \times 10 = 450$ (add 1 zero to 45)

(2) $67 \times 100 = 6700$ (add 2 zeros to 67)

(3) $65 \times 1000 = 65000$ (add 3 zeros to 65)

(4) $65788 \times 1000000 = 65788000000$ (add 6 zeros to 65788)

Speed Method: Rules for numbers with decimals:

When multiplying by 10, 100, 1000 move the decimal place the appropriate number of places to the right.

(1) $67.5 \times 10 = 675$ (the decimal point is moved 1 place to the right to give us 675.0 which is the same as 675)

(2) $67.5 \times 100 = 6750$ (this time move the decimal point two places to the right to give 6750.0 which is the same as 6750)

(3) $6.87 \times 1000 = 6870$ (in this case move the decimal point three places to the right to give the required answer.)

Now consider examples involving division by 10, 100 and 1000.

(1) $450 \div 10 = 45$ (You simply remove one zero from the number)

(2) $5600 \div 100 = 56$ (This time you remove two zeros from the number)

(3) $45 \div 100 = 0.45$ (No zeros to remove – so this time move the decimal point two places to the left to give us 0.45)

(4) $345.78 \div 100 = 3.4578$ (Again simply move the decimal point 2 places to the left to give the answer)

(5) $456.78 \div 1000 = 0.45678$ (Move the decimal point 3 places to the left as shown)

(6) $458 \div 0.1 = 4580$ (remember 0.1 means one–tenth, so dividing a number by 0.1 or one-tenth means the answer becomes 10 times bigger.)

Mental arithmetic questions involving powers of 10

(1) Divide 27000 Milliliters by 100

(2) What is78.87 multiplied by 1000?

(3) What is 67 divided by 100?

(4) What is 687 divided by 0.1?

Using the methods shown earlier the answers are:

(1) 270 ml (2) 78870 (3) 0.67 (4) 6870

If you feel comfortable with the methods above you can skip the traditional method below - although if you have time it will add to your conceptual understanding and will help explain why the 'speed method' leads to the correct answers

Traditional method of multiplying by 10

The traditional method of multiplying by a 10, 100, 1000 is shown below. This method is useful as it cements the conceptual understanding required. Consider having to work out 34×10

Consider place value. For example for the number 34, the right hand digit is the units digit and the number 3 on the left hand side is the tens digit or column. In fact every time you move one place to the left you increase the value by 10. So moving left by one place from the tens column we get the 100's column as shown below.

Hundreds	Tens	Units
	3	4

When we multiply by 10 each digit moves one column to the left. So, $34 \times 10 = 340$ as shown below. In other words 3 tens becomes 3 hundreds, the 4 units becomes 4 tens as shown. Also notice we have 0 units so we must put a zero in the units column. Moving each digit 1 place to the left has the effect of making it 10 X bigger.

Hundreds	Tens	Units
3	4	0

Consider the sum 34×100

Multiplying by 100 is similar. We simply multiply by 10 and then 10 again. This has the effect of moving each digit two places to the left. This makes it $100 \times$ bigger.

The number 34 is shown below as 3 tens and 4 units.

Thousands	Hundreds	Tens	Units
		3	4

We will now do the multiplication and see its effect.

Clearly, multiplying 34 by 100 has the effect of moving the 3 in the tens column to the thousands column and the 4 units to the hundreds column. This is shown below.

Thousands	Hundreds	Tens	Units
3	4	0	0

So $34 \times 100 = 3400$ as shown above.

This technique is important as it illustrates the concept of multiplying by 10 or 100 taking place. The same process applies to multiplying by 1000, 10,000 or a higher power of 10.

Also note, there is a short hand way of writing 100, 1000, 10,000 and larger powers of 10.

$100 = 10^2$ (10 squared, which is 10×10)

$1000 = 10^3$ (10 cubed which is $10 \times 10 \times 10$)

$10,000 = 10^4$ ((10 to the power 4, which is $10 \times 10 \times 10 \times 10$)

$1000,000 = 10^6$ (10 to the power 6 which is $10 \times 10 \times 10 \times 10 \times 10 \times 10$)

Higher powers can be written similarly.

Dividing by 10, 100 and 1000.

Conceptually, dividing by 10, 100 or 1000 is a similar process, except, on this occasion, you move the digits to the right by the appropriate number of places.

Consider having to divide 34 by 10.

Here, 3 tens and 4 units becomes 3 units and 4 tenths as shown.

Hundreds	Tens	Units	Tenths
		3	4

The rationale for this is that we move each digit to the right. So 3 tens becomes 3 units and 4 units becomes 4 tenths as shown above. The answer is written as 3.4. Similarly, when dividing by 100 or a 1000 the number is moved two and three places to the right as appropriate. **We saw earlier the technique to work out the answer mechanically. This ensures you get the right answer without having to resort to the thousands, hundreds, tens, units, tenths and hundredths column. The simple rules shown on pages 12 and 13 may help those students who find the above process difficult.**

Chapter 2 Mental Arithmetic Part 2

Most questions in the Mental Arithmetic test will require several steps and include various operations i.e. **+, - , x and ÷**

Time Based Questions

(QTS questions seem to require answers in 24 hour clock)

For converting time from 12 hour clock to 24 hour clock see examples below

12 –Hour Clock	24 –Hour Clock
8.45 am	08:45
11.30 am	11:30
12.20pm	12:20
2.35 pm	14: 35 (after 12pm add the appropriate minutes and hours to 12 hours, in this case 2hrs 35mins +12hrs = 14:35)
8.45 pm	20:45 (8hrs 45mins + 12hrs = 20:45)
11.47pm	23:47 (11hrs 47mins +12hrs = 23:47)

The convention is that if the time is in 24-hr clock there is no need to put hrs after the time.

Also remember:

2.5 hours = 2 hours 30minutes (0.5 hours = half of 60 minutes)

2.4 hours = 2 hours 24 minutes (0.4 hours = 0.4×60 = 24 minutes)

For other time based questions e.g. years, months, days, hours, minutes or seconds remember the appropriate units.

Example 1: At a parents evening a teacher has to see the parents of each pupil for 12 minutes. There are 15 pupils. Also there is a break of 20 minutes. The session starts at 5.30pm. When does it finish? Give your answer using the 24 hour clock

Method: Clearly we need to first work out the total time it takes for all the pupils including the break time. Total time for 15 pupils is $15 \times 12 = (15\times10 + 15\times2) = 180$ minutes = 3 hours plus break time of 20 minutes. So the parents evening stops 3hrs and 20 minutes after 5.30pm – this means it ends at 8.50pm. However using the 24 hour clock the times it ends is 20:50

Example 2: At a junior school a child completes a lap in 2.4 minutes. How many minutes and seconds is this?

Convert 0.4 minutes into seconds. Since one whole minute = 60 seconds, then 0.4 minutes = $0.4\times60 = 24$ seconds. Hence the child completes the lap in 2 minutes and 24 seconds.

(Note that 0.4×60 is the same as 4×6, since if you multiply $0.4 \times 10 = 4$, correspondingly divide 60 by 10 to get 6)

Example 3

Example of reading age

A child's actual age is 8 years and 11 months. However, her recent test report indicates that her reading age is 6 months above her actual age. What is the pupil's reading age in years and months?

Method: So we have to add 6 months to 8 years and 11 months

=9years, 5months

General Multiplication questions

Example 1: There are 4 classes of 18 children and 3 classes of 23 children going for a school outing. How many pupils are there altogether?

Method: 4 classes of 18 imply there are 4×18 pupils $= 72$

(Another way of working out 4×18 is to break it down as follows: $4 \times 18 = 4 \times 10 + 4 \times 8 = 40 + 32 = 72$)

Similarly, 3 classes of 23 means, $3 \times 23 = 69$ pupils

Finally, $72 + 69 = 70 + 2 + 60 + 9 = 130 + 11 = 141$

There are a total of 141 pupils.

Example 2:

I buy 5 books for £3.97 each. How much change do I get from a £20 note?

Method: Round up each book to £4. Hence the cost of 5 books = £4×5 – 5×3p = £20 - 15p = £19. 85

You can see straight away that I get 15p change from my £20 note

More multiplication methods that may be helpful

The Grid Method of Multiplication

This is a very powerful method for those who find traditional long multiplication methods difficult.

Example 1: Multiply 37×6

Re-write the number 37 as 30 and 7 and re-write as shown in the grid table.

×	30	7
6	180	42

Now simply add up all the numbers inside the grid. So the answer is 180+42 =222

Example 2: work out 15×13

To work this out using the grid method, re-write 15 as 10 and 5, and 13 as 10 and 3 as shown on the outside of the grid table.

×	10	5
10	100	50
3	30	15

Multiply out the outside horizontal numbers with the outside vertical numbers to get the numbers inside as shown. Finally, just add up the inside numbers which in this case is 100+50+30+15 =195

Multiplication with decimals

Example 3: Work out 1.5×1.3

Step1: Leave out the decimal points and just work out the answer to 15×13 as shown above.

We know the answer to this is 195.

Step2: Now count the number of digits there are from the right before the decimal point for each number being multiplied and add them up. That is one for the first number and one for the second number to give a total of 2.

Step3: In the answer 195 count two from the right hand side and insert the decimal point.

So the answer is 1.95

Example 4: Work out 0.15×1.3

We know the answer to 15×13 is 195

Again counting from right, this time the number of digits for each number before the decimal point is 2 for the first number and 1 for the second number giving a total of 3. We now count 3 places from the right and insert a decimal point.

So the answer is 0.195

If you want to you can think of getting the answer another way:

Consider **Example 3** again: Multiply $1.5\times .3$

We know the answer is 195. Note the fact that 1.5 is 15 divided by 10 and 1.3 is 13 divided by 10. So the answer is simply 195 divided by $10\times10 = 100$, so we divide 195 by 100 to get the answer as 1.95

More Multiplication

We will look at some fascinating ways of quickly multiplying by 11, 9, and 5, which will help you speed up your number work in mental arithmetic

Multiplying quickly by 11

One common method used is to multiply by 10 and then add the number itself. We will now look at a super- efficient method that is rarely used.

Super-efficient Speed Method:

$11 \times 11 = 121$ (the first and last digits remain the same & the middle number is the sum of the first two digits)

The basic method is: Start with the first digit, add the next two, until the last one. This method works with any number of digits.

Let us explore a few more examples with two digit numbers.

$13 \times 11 = 143$ (Keep the first and last digit of the number 13 the same, add 1 & 3 to give the middle number 4)

$14 \times 11 = 154$

$19 \times 11 = 1(10)9 = 209$ (Notice the middle number is 10, since $1+9=10$, so we need to carry 1 to the left hand number.)

A few more examples will show the power of this method.

$27 \times 11 = 297$ (the first number=2, the middle number=2+7, the last number =7)

$28 \times 11 = 2(10)8 = 308$ (using similar analysis to 19×11 above)

The same principle applies to numbers with more than 2 digits.

Example: Work out 215×11

Method: Keep the first and the last digit the same. Starting from the first digit add the subsequent digit to get the next digit, do this again with the second digit until the last digit which stays the same. So, $215 \times 11 = 2365$ (2, is the first digit so stays the same, the sum of 2 and 1 gives you the next digit 3, the sum of 1 and 5 gives you the third digit 6 and finally the last digit 5 stays the same)

Example involving multiplying by 11

There are 54 pupils in year 7 who collect 11 merits each in one month.

How many merits in total did the 54 pupils receive?

54×11 using the method explained above is 594

Method: Keep the first and last digit of the number 54 he same, add 5 & 4 to give the middle number 9

Multiplying quickly by 9

Here is an easy method to work out the $9\times$ table

Example 1: Work out 9×7

Method:

Step1: Add '0' to the number you are going to multiply by 9, e.g. 7 to get 70

Step2: Now subtract 7 from 70 to get 63 which is the final answer

Example 2: Work out 9×12

Method:

Step1: Add '0' to the number you are going to multiply by 9, i.e. 12 to get 120,

Step2: Now subtract 12 from 120 to get 108 which is the final answer

Example 3: Work out 9×35

Method:

Step1: Add '0' to the number you are going to multiply by 9, e.g. 35 to get 350,
Step2: Now subtract 35 from 350 to get 315 which is the final answer

Essential Method: $6 \text{ X } 9 = 6(10 - 1) = 60 - 6 = 54$

A quick way of multiplying by 5

Multiply the number by 10 and halve the answer.

Example 1: $5 \times 4 =$ half of $10 \times 4 =$ half of $40 = 20$

Example 2: $5 \times 16 =$ half of $10 \times 16 =$ half of $160 = 80$

Example 3: 5 × 23 = half of 10 × 23 = half of 230 = 115

TIP: Remember the Order of Arithmetical Operations

Remembering the order in which you do arithmetical operations is very important .The rule taught traditionally is that of **BIDMAS.**

The **BIDMAS** rule is as follows:

(1) Always work out the **B**racket(s) first

(2) Then work out the **I**ndices of a number (squares, cubes, square roots and so on)

(3) Now **M**ultiply and **D**ivide

(4) Finally do the **A**ddition and **S**ubtraction.

Example 1: Work out 2 + 8×3

Do the multiplication before the addition

So 8×3 =24 then add 2 to get 26

Example 2: 4 + 13(7 – 2) this means add 4 to 13×(7 – 2)

Do the **brackets first** so 7 – 2 =5, **then multiply** 5 by 13 to get 65 and **finally add** 4 to get 69

Example 3: work out 3^2 × 5 – 9

Work out the **square of 3 first**, then **multiply by 5** and finally **subtract 9** from the result.

So we have 3×3 = 9, 9×5 = 45 and finally 45 - 9 = 36

Summary

When working out sums involving mixed operations (e.g. +, - , x and ÷) you need to work out the steps in stages using the BIDMAS rule

So to work out 8 +25 ×12

Do the multiplication first, 25×12 =300, write down 300 then add 8 to get the answer 308.

Division

In general, the traditional short division approach is a good method. However, there are some other smart techniques worth considering for special situations.

Dividing a number by 2 is a very useful skill, since if you can divide by 2, you can by halving it again divide by 4 and halving it again divide by 8.

Dividing by 2, 4 and 8

Simply halve the number to divide by 2

(Some find it difficult to halve a number like 13. An alternative strategy is to multiply the number by 5 and divide by 10)

Halving again is the same as dividing by 4

And halving once more is the same as dividing by 8

Example 1: $28 \div 2 = 14$

Example 2: $268 \div 4 = 134 \div 2 = 67$

Example 3: $568 \div 8 = 284 \div 4 = 142 \div 2 = 71$

Example 4: $65 \div 4 = 32.5 \div 2 = 16.25$

Dividing by 5

An easy way to do this is to multiply the number by 2 and divide by 10.

Example 1: $120 \div 5 = (120 \times 2) \div 10 = 240 \div 10 = 24$

Example 2: $127 \div 5 = (127 \times 2) \div 10 = 254 \div 10 = 25.4$

Similarly to divide by 50 simply multiply by 2 and divide by 100

Dividing by 25

A good way to do this is to multiply by 4 and divide by 100.

Example 1: $240 \div 25 = (240 \times 4) \div 100 = 960 \div 100 = 9.6$

Example 2: $700 \div 25 = (700 \times 4) \div 100 = 2800 \div 100 = 28$

Dividing by other numbers: The conventional short division method is a good method but you might find the speed methods below useful sometimes.

Mental Arithmetic Question involving division

Example: £67.5 is divided amongst three pupils. How much does each pupil get?

Clearly this is the same as $60 \div 3$ added to $7.5 \div 3$

$60 \div 3 = 20$ and $7.5 \div 3 = 2.5$ which altogether is 22.5

Hence, £$67.5 \div 3 =$ £22.50 per pupil

Example 1:

Divide 145 by 7

(145 = 140 +5)

We can say that 140 ÷ 7 =20, and then we are left with 5/7.

So the answer is 20 and 5/7

Example 2:

Divide 103 ÷ 9

(103 =99 + 4)

=99 ÷ 9 + 4/9

= 11 and 4/9

Rounding numbers and estimating

We will start simply with rounding numbers to the nearest 10 and 100

Consider the number 271

Rounded to the nearest 10 this number is 270

Rounded to the nearest 100 this number is 300

(The principle is that if the right-hand digit is lower than 5 you drop this number and replace it by 0. Conversely if the number is 5 or more drop that digit and add 1 to the left)

Try a few more:

5382 to the nearest 10 is 5380

5382 to the nearest hundred is 5400

5382 to the nearest 1000 is 5000

This rule can also be applied to decimal numbers:

3.7653 rounded to the nearest thousandth is 3.765

3.7653 rounded to the nearest hundredth is 3.77

3.7653 rounded to the nearest tenth is 3.8

3.7653 rounded to the nearest unit is 4

Tip: remember to use common sense when rounding in real life situations:

Example: A teacher wants to keep 120 English text books in the same size boxes. She can fit 22 text books in a box. How many boxes will she need?

Method: Number of boxes required will be 120÷22= 5.5 (to one decimal place). But clearly, she cannot have 5.5 boxes. So she needs to have 6 boxes

Estimating calculations quickly

Example 1: Work out $(2.2 \times 7.12)/4.12$

We can quickly estimate that this is roughly equal to $(2 \times 7)/4$ $=14/4$ which is around 3.5 or 4 rounded to the nearest unit.

The actual answer is: 3.8 (to 1 decimal place)

Example 2: Work out $38 \times 2.9 \times 0.53$

We can approximate 38 to be 40 to the nearest ten

We can approximate 2.9 o 3 to the nearest unit

We can approximate 0.53 to 0.5 to the nearest tenth

So the magnitude of the answer is $40 \times 3 \times 0.5$

This is $120 \times 0.5 = 60$ (approximately)

Chapter 3 General Arithmetic Part 1

Fractions, decimals and percentage equivalents

I am sure most of you are aware that $\frac{1}{2}$ =0.5. This in turn is equal to 50%.

It is worth reviewing this fact. In addition, you should try and remember the following other equivalences if you have forgotten them:

Fractions, decimals and percentage equivalents

Fractions	Decimal	Percentage
$\frac{1}{2}$	**0.5**	**50%**
$\frac{1}{4}$	**0.25**	**25%**
$\frac{3}{4}$	**0.75**	**75%**
$\frac{1}{10}$	**0.1**	**10%**
$\frac{1}{5}$	**0.2**	**20%**

If, we know $\frac{1}{2}$ **= 0.5**

We can deduce that $\frac{1}{4}$ = **0.25**

(Since a quarter is half of half)

Similarly, $\frac{1}{8}$ is **0.125**

We can do this quickly because all we do is halve each decimal value.

Half of 0.5 is 0.25, Half of 0.25 is 0.125

We can of course continue this process.

Further if we know $\frac{1}{10}$ =0.1 we can now work out $\frac{2}{10}, \frac{3}{10}, \frac{7}{10}$ etc.

$\frac{2}{10}$ **= 0.2 (2 × 0.1),** $\frac{3}{10}$ **= 0.3 (3 × 0.1,** $\frac{7}{10}$ **= 0.7 (7 × 0.1),** $\frac{9}{10}$ **= 0.9 (9 × 0.1)**

Another useful fraction and decimal equivalent to remember is $\frac{1}{3}$

=0.333… (0.3 recurring)

The key equivalent percentages to remember are as follows:

$\frac{3}{4}$ **= 75%,** $\frac{1}{2}$ **=50%,** $\frac{1}{4}$ **= 25%,** $\frac{1}{8}$ **= 12.5%,** $\frac{1}{10}$ **= 10%**

See summary box below

Summary:

Remember the following equivalences

$\frac{1}{2}$ **=0.5 =50%,** $\frac{1}{4}$ **=0.25 =25%,** $\frac{3}{4}$ **=0.75 =75%,** $\frac{1}{10}$ **=0.1=10%**

Also if you can try to remember, $\frac{1}{5}$ **=0 .2 = 20%, and** $\frac{2}{5}$ **=0.4 =40%,** $\frac{1}{3}$

=0.333… (0.3 recurring) =33.33% (to 2 decimal places)

To convert a fraction into a percentage, simply multiply the fraction by 100

Mental Arithmetic questions involving percentages and fractions

Example 1: Find 25% of £250

Method: Find 50% of £250 and halve it again.

Half of £250 = £125, Half of £125 =£62.50, so 25% of £250 = £62.50

Example 2: In a class of 25 pupils there are 12 girls and the rest are boys.
(1) What fraction consists of boys? (2) What percentage is this?

Method:

(1) Since there are 12 girls, there are 13 boys out of 25. So the fraction of boys is $\frac{13}{25}$

(2) The percentage of boys is $\frac{13}{25}$ X 100 = 52%, (Divide 100 by 25 to get 4. Then multiply 13 by 4 to get 52%)

Example 3: 30% of the pupils in a class are boys. There are 30 pupils altogether. How many pupils are girls in this class?

Method: If 30% of the pupils in a class are boys, clearly 70% are girls. So we need to find 70% of 30 pupils. Since 10% of 30 is 3, this means 70% corresponds to 3X7 = 21 girls. Hence, this class has 21 pupils that are girls.

Working out increase or decrease in percentages from original value

Example 1: In a certain school 16 pupils got a grade B in Science in 2010. In the same school 20 pupils got a grade B in Science in 2011. What was the percentage increase in grade B's from 2010 to 2011 in this subject?

Method: Increase in number of grade B's = 20 – 16 =4. Original number of pupils =16. The increase of 4 was based on 16 pupils. To work out the percentage increase we simply divide the increase by the original number of pupils with grade B and multiply this by 100. That is $\frac{4}{16}\times 100 = \frac{1}{4} \times 100 = 25\%$

To work out decrease in percentages (uses the same principle as above)

Example 2: The original price of a classroom projector was: £150, the new price is reduced to £135. What is the percentage decrease in price? The decrease in price is £150 - £135 = £15. The decrease over the original price is $\frac{15}{150}$, to turn this into a percentage we multiply $\frac{15}{150} \times 100 = \frac{1500}{150} = 10\%$. So the

decrease in percentage price is 10%. The basic formula to work out increase or decrease percentage change is shown below:

$$\frac{\textit{difference between final and original value}}{\textit{original value}} \times 100$$

One thing to remember though is that the increase or decrease in percentage points is different from **increase or decrease in percentages**.

To illustrate this consider the example below:

The unemployment rate in a region A was 8% in 2010. In 2011 the unemployment rate in the same region was 10%. **(1)** What was the **percentage point** increase in unemployment from 2010 to2011? **(2)** What is the **percentage increase** in unemployment from 2010 to 2011?

(1) The **percentage point** increase is simply 2% (i.e. from 8% to 10%)

(2) However the **percentage increase** in unemployment is $\frac{2}{8} \times 100 = \frac{200}{8} = \frac{100}{4}$

= 25%. In the QTS Numeracy Skills test context, if you are asked to work out the **percentage point increase,** say in GCSE success rates from 20% to 30%. The answer is obviously 10%. But if asked to work out the **percentage increase**, then the answer is $\frac{10}{20} \times 100 = \frac{1000}{20} = \frac{100}{2} = 50\%$

Miscellaneous questions involving fractions and percentages

Example 1: Finding fraction of an amount

Find $\frac{3}{4}$ of £600

First find $\frac{1}{2}$ = £300

Then find $\frac{1}{4}$ (which is half of half) = £150

Therefore $\frac{3}{4}$ = £450 (adding half plus a quarter)

Example 2: Finding a fraction and turning it into a percentage

There are 80 pupils in a rural junior school. 10 pupils need additional help with reading.

What is the percentage of pupils that need reading help?

The fraction of pupils that need help $= \frac{10}{80}$, by dividing top and bottom numbers by 10 we get $\frac{1}{8}$

To convert 1/8 into a percentage simply multiply 1/8 by 100

= 1/8X100 =100/8 = 50/4 =25/2 =12.5%

(Another method: We know ¼ =25%

Hence 1/8 =12.5% (Since 1/8 is half of a quarter)

Calculator based questions

You need to remember that percent means out of 100. That is $\frac{1}{100}$. So to find say 42.5% of a number, divide the number by 100 and multiply it by 42.5.

Example: work out 42.5% of £400

We can say that this is the same as (£400 ÷ 100) X 42.5 =4 × 42.5 = £170

Try working this out with a basic calculator to see if you agree with the answer.

When using the On screen calculator. Do the calculations in steps. For example to work out 3.65 + 15X6

Use the rule that you always do multiplication and division before you do addition and subtraction. So to work out 3.65 +15×6, we work out 15×6 first. This gives us 90, then we add 3.65 to 90 to give us 93.65 as the final answer. Note to clear the answer simply click on [C] on the calculator. For interest the [CE] key is used to clear only the last entry. As stated below do the calculations in steps to make sure you get the correct answer.

Simplifying fractions

Reducing a fraction to its lowest terms

Basically you need to find numbers that divide into the top number (numerator) as well as the bottom number (denominator), and then divide them both by the same number (start with 2, if doesn't go then choose 3, then 5, and then the next prime factor e.g. 7, 11, etc.)

Example1: Reduce $\frac{16}{24}$ to its lowest terms.

8 divides exactly into 16 and 24, so in the fraction $\frac{16}{24}$ divide top and bottom by

8. This gives the answer as $\frac{2}{3}$

In case you can't see this straight away, try starting with the number two and work your way numerically upwards using the next prime factor i.e. try 3, then 5 etc. if required

So for the fraction $\frac{16}{24}$ we can start dividing top and bottom by 2 to give us $\frac{8}{12}$,

then do the same again as both 8 and 12 are still divisible by 2. This gives us $\frac{4}{6}$

and finally repeating the process once more reduces the fraction to $\frac{2}{3}$ which is

the simplest form.

Example 2: Simplify $\frac{9}{12}$ to its lowest terms. In this case we can't divide top and

bottom by 2, so we try 3. Since 3 will go into both 9 and 12, we can reduce this to the fraction $\frac{3}{4}$ (since 9 ÷3 =3 and 12 ÷ 3 =4) Hence, $\frac{9}{12}$ reduces to $\frac{3}{4}$

Example 3:

Reduce fraction $\frac{49}{77}$ to its lowest terms. This time we need to spot that 2, 3, 5,

does not go into either 49, or 77. Either by trial and error or by spotting the right number we notice 7 goes into both the numerator and the denominator. This reduces

$\frac{49}{77}$ to $\frac{7}{11}$

Cancelling down fractions to their simplest form (lowest terms)
To simplify a fraction to its lowest terms you divide the numerator and the denominator by the same prime factors (2, 3, 5, 7, 11, etc.) to give the equivalent fractions as shown in the examples above

Finding fraction of an amount

Example: Find $\frac{2}{5}$ of 25

Simply replace the 'of' by ×. (times)

So $\frac{2}{5}$ of 25 becomes $\frac{2}{5} \times 25$

To work this out find out 1/5 of 25 and then multiply the answer by 2.
So 25 divided by 5, equals 5, then 2 x 5 =10

Hence $\frac{2}{5}$ of 25 =10

Example: On screen based question

There are 60 pupils altogether in a year 10 group. 12 pupils are entered for GCSE maths in that year. What is the proportion of the pupils that are not entered for maths? Give your answer as a decimal.

Total number of pupils = 60, pupils entered for GCSE maths = 12, this means 48 are not entered for maths. Hence the proportion that is not entered = 48/60. If you divide top and bottom by 6, this simplifies to 8/10.

The answer as a decimal this is 0.8

Adding and Subtracting Fractions

This next section will help you revise adding, subtracting, multiplying and dividing fractions together.

Consider adding and subtracting fractions together.

When the bottom numbers (denominators) are the same, just add the top numbers together keeping the bottom number the same. Likewise for subtraction just subtract the top two numbers.

Example 1: $\frac{2}{5} + \frac{1}{5} = \frac{3}{5}$

Example 2: $\frac{2}{5} - \frac{1}{5} = \frac{1}{5}$

When the denominators are different

Example3: Work out $\frac{1}{2} + \frac{2}{5}$

When the denominators are different, the traditional method of doing this is to find the lowest common denominator. We have to find a number that both 2 and 5 will go into. This is clearly 10.

We can now re-write the fraction with the same common denominator.

To do this we have to ask how did we get the denominator from 2 to 10 for the first part, and likewise for the second part from 5 to 10. The answer is shown below:

$$\frac{1X5}{2X5} + \frac{2X2}{5X2} = \frac{5}{10} + \frac{4}{10} = \frac{9}{10}$$

We had to multiply top and bottom by 5 for the first part and top and bottom by 2 for the second part as shown above. We can then add the fraction as we have the same common denominator.

We can however use another very simple strategy that always works. The method is that of crosswise multiplication.

The basic method is to take the fraction sum and do crosswise multiplication as shown by the arrows. In addition, multiply the denominators (bottom numbers) together to get the new denominator.

Example1: $\frac{1}{2}\frac{1}{2} \times \frac{2}{5} = \frac{1}{2} + \frac{2}{5} = \frac{1X5+2X2}{2X5} =$

$\frac{5+4}{10} = \frac{9}{10}$

We notice that if we cross multiply as shown we get 1×5 and 2×2 respectively at the top. To get the bottom number we simply multiply the bottom numbers, 2 and 5 together. So the denominator is 2 X 5=10.

Let us try another example:

Example2: Work out $\frac{3}{7} + \frac{2}{5}$

Using crosswise multiplication and adding rule, as well as multiplying the bottom two numbers we get:

$$\frac{3}{7} \times \frac{2}{5} = \frac{3X5+7X2}{35} = \frac{15+14}{35} = \frac{29}{35}$$

This is a very elegant method which always works

Example3: Work out $\frac{3}{7} - \frac{2}{5}$

This is similar to the above except instead of adding we now subtract as shown below.

$$\frac{3}{7} \times \frac{2}{5} = \frac{3X5-7X2}{35} = \frac{15-14}{35} = \frac{1}{35}$$

Note: In fact you can use this method when adding or subtracting any fraction that you find difficult. Even if you use this method for simple cases, you will still get the right answer but you may have to cancel down to get the lowest terms for the final answer.

For example we know that $\frac{1}{4} + \frac{1}{2} = \frac{3}{4}$

But if we didn't know and used the method shown we would get $\frac{1}{4} \times \frac{1}{2}$

$= \frac{1X2+4X1}{4X2} = \frac{2+4}{8} = \frac{6}{8} = \frac{3}{4}$ (we get this by dividing both the

numerator and denominator in $\frac{6}{8}$ by 2). So we get the same answer in the end

Mental Arithmetic questions involving fractions.

(1) Find $2\frac{3}{4}$ of 64

Step 1: We first work out 2 X 64 = 128

Step 2: To work out three quarters of 64, we first work out a half and then add it to a quarter of 64.

Half of 64 is 32

A quarter of 64 (is half of 32) is 16

Step 3: Hence three quarters of 64 = 32 + 16 = 48

Step 4: So two and three quarters of 64 = 128 + 48 = 176

Adding and subtracting mixed numbers

This is a similar process. We first add or subtract the whole numbers and then the fractional parts.

Example 1: $2\frac{2}{5} + 4\frac{3}{7}$

Adding the whole numbers we get 6. (Simply add 2 and 4)

Now add the fractional parts to get: $\frac{14+15}{35} = \frac{29}{35}$

So the answer is $6\frac{29}{35}$

Example 2: $4\frac{3}{7} - 2\frac{2}{5}$

Subtract the whole numbers and then the fractional parts, which gives us:

$$2\frac{15-14}{35} = 2\,\frac{1}{35}$$

Multiplying Fractions

Multiplying fractions by the traditional method is quite efficient so we will consider only this approach.

Example 1: $\frac{2}{3} \text{ X } \frac{5}{7} = \frac{10}{21}$

In this case we simply multiply the top two numbers to get the new numerator and multiply the bottom two numbers together to get the new denominator, as shown above.

Another example will help consolidate this process:

Example 2: $\frac{10}{21} \text{ X } \frac{5}{7} = \frac{50}{147}$

(Multiply 10 X 5 to get 50 for the numerator and 21 X 7 to get 147 for the denominator)

Division of Fractions

When dividing fractions we invert the second fraction and multiply as shown.

Think of an obvious example. If we have to divide ½ by ¼ we intuitively know that the answer is 2. The reason for this is that there are 2 quarters in one half. Let us see how this works in practice.

Example 1: $\frac{1}{2} \div \frac{1}{4} = \quad \frac{1}{2} \text{X} \frac{4}{1} = \frac{4}{2} = 2$

Step 1: Re-write fraction as a multiplication sum with the second fraction inverted.

Step 2: Work out the fraction as a normal multiplication

Step 3: Simplify if possible. In this case 4 divided by 2 is 2.

Example 2: $\frac{6}{11} \div \frac{5}{11} = \frac{6}{11} X \frac{11}{5} = \frac{66}{55} = \frac{6}{5} = 1\frac{1}{5}$

Step1: Re-write the fraction inverting the second fraction as shown

Step2: Multiply the top part and the bottom part to get $\frac{66}{55}$ **as shown.**

Step 3: Simplify this by dividing top and bottom by 11 to get $\frac{6}{5}$**. Now this finally simplifies to** $1\frac{1}{5}$ **as shown.**

The following steps are required to convert a mixed number into a fraction.

Step 1: Multiply the denominator of the fractional part by the whole number and add the numerator. Consider the mixed number $2\frac{1}{4}$. This works out to $2 \times 4 + 1 = 9$. This now becomes the new numerator.

Step 2: The new denominator stays the same as before. Now re-write the new fraction as $\frac{9}{4}$. (That is the new numerator $\div$ existing denominator)

Example 3: Convert the mixed number, $3\frac{3}{7}$ into a fraction.

Step 1: Multiply denominator of fractional part by whole number and add the numerator. This gives $3 \times 7+3 = 24$ as the new numerator. **Step2:** Re-write fraction as new fraction. This is now the new numerator $\div$ existing denominator. This gives us $\frac{24}{7}$

Multiplying mixed numbers together

Consider the examples below:

Example: $1\frac{1}{5} \times 1\frac{3}{8}$

The method is simply to convert both mixed numbers into fractions and multiply as shown below:

$$1\frac{1}{5} x 1\frac{3}{8} = \frac{6}{5} x \frac{11}{8} = \frac{66}{40} = 1\frac{26}{40} = 1\frac{13}{20}$$

Notice $\frac{26}{40}$ simplifies to $\frac{13}{20}$

Dividing mixed numbers together

Example: $1\frac{1}{2} \div 1\frac{1}{4}$

There are two steps required to work out the division of mixed numbers.

Step1: Convert both mixed numbers into fractions as before

Step 2: Multiply the fractions together but invert the second one.

$$1\frac{1}{2} \div 1\frac{1}{4} = \frac{3}{2} \div \frac{5}{4} = \frac{3}{2} \text{ x } \frac{4}{5} = \frac{12}{10} = 1\frac{2}{10} = 1\frac{1}{5}$$

Chapter 4 General Arithmetic Part 2

Proportions and ratio

Although proportion and ratio are related they are not the same thing – see example below for clarification.

Example: In a class there are 15 girls and 10 boys. The **ratio of girls to boys is** 15:10, or 3:2, (divide both 15 and 10 by 5) and the **proportion of girls in the class** is 15 out of 25 or $\frac{15}{25}$ **which simplifies to** $\frac{3}{5}$

Mental Arithmetic questions based on proportions and ratios

Example 1:

In a class of 27 pupils, 9 go home for lunch. What is the proportion of pupils in this class that have lunch at school?

Since 9 out of 27 pupils go home, this means 18 pupils have lunch at school.

As a proportion this is 18 out of 27 or $\frac{18}{27}$ which simplifies to $\frac{2}{3}$

Example 2: The ratio of boys to girls in a class is 2: 3. There are 25 pupils altogether. How many boys are there?

Step 1: Find out the total number of parts, you can do this by adding up the ratio parts together. E.g. 2:3 means there are (2+3) = 5 parts altogether. This means 1 part = one fifth of 25 pupils = 5 pupils.

Since the ratio of boys to girls is 2:3, there are 2×5 boys and 3×5 girls

The number of boys in the class =2×5 = 10

Scales and ratios

Consider that you are reading a map and the scale ratio is 1: 100000

This means for every one cm on the map the actual distance is 100000 cm or put another way every one cm on the map, the distance = 1000 m (divide 100000 by 100)

to get the result in metres). Now, 1000 m = 1km (divide 1000 by 1000 to get 1 since 1km =1000m)

(Scales can also be used in other areas such as architectural drawings)

Mental Arithmetic question based on scales

I note that the map I am using has a scale of 1: 25000. The distance between the two places I am interested in is 12cm. What is the actual distance in km?

Method: 12 cm on the map corresponds to 12 x 25000 = 300 x 1000 =300000cm

=3000m = 3km

(300000 /100 to convert to metres = 3000 m, now divide 3000 by 1000 to convert to km)

Hence the distance between the two places is 3km

Conversions

Conversions are often useful in changing currencies for example from pounds to dollars or euros or vice-versa. It is also useful to convert distances from miles to kilometers or weights from kilograms to pounds and so on.

Basically a conversion involves changing information from one unit of measurement to another. Consider some examples below:

On Screen based question based on conversions

Example 1:

I go to France with £150 and convert this into Euros at 1.2 Euros to a pound. **(1)** How many Euros do I get? **(2)** I am left with 39 Euros when I get back home. The exchange rate remains the same. How many pounds do I get back?

Method: (1) Since 1 pound = 1.2 Euros, I get 150 × 1.2 =180 Euros in total.

Method: (2) When I get back I change 39 Euros back into pounds. This time I need to divide 39 by 1.2

So 39÷1.2 =32.5. This means I get back £32.50

Example 2:

The formula for changing kilometers to miles is given by, $\mathbf{M} = \frac{5}{8} \text{ X } \mathbf{K}$

Use this formula to convert 68 kilometers to miles

Method: substitute **K** with 68 and multiply by $\frac{5}{8}$

This means $M = \frac{5}{8} \text{ X } 68$. Using a calculator this comes to 42.5 miles

It is worth reviewing some common Metric and Imperial Measures as shown below

Metric Measures

1000 Millilitres (ml) =1 Litre(l)

100 Centilitres (cl) =1 Litre (l)

10ml =1 cl

1 Centimetre (cm) =10 Millimetres (mm)

1 Metre (m) = 100 cm

1 Kilometre (km) =1000 m

1 Kilogram (kg) =1000 grams (g)

Imperial Measurements

1 foot =12 inches

1 yard =3 feet

1 pound = 16 ounces

1 stone =14 pounds (lb)

1 gallon = 8 pints

1 inch = 2.54 cm (approximately)

On Screen based question on conversions

A teacher takes some of her sixth form students on a walking tour whilst in the South of France. They walk from Perpignan to Canet Plage which is approximately 11 km away. After a lunch break and some time on the beach, they walk back to Perpignan. How many miles in total do they walk on that day? (You are given that 8 km is approximately equal to 5 miles.) Give your answer as a decimal.

Method: Total distance walked = 22Km (11 + 11) To convert this into miles we have to multiply 22 by 5 and then divide by 8 (Since 8 km = 5 miles) That is $22\times\frac{5}{8}$ =13.75 miles (simply multiply 22 by 5 and divide the answer by 8)

(Remember, use of calculator is allowed for all On screen questions)

Chapter 5 General Arithmetic part 3

Weighted Average

When two or more sets of data are combined together and some are more important than others, then different weightings are given. The weighted mean is then worked out appropriately as shown in the examples below:

Example 1:

In a particular subject the coursework carries a weight of 0.3 and the exam mark carries a weight of 0.7

Calculate the overall percentage result if a pupil got 60% for the coursework and 50% for the exam.

The percentage result is $60 \times 0.3 + 50 \times 0.7 = 18 + 35 = 53\%$

Example 2

A pupil achieved the following marks in three tests

Test 1: 55 marks

Test 2: 60 marks

Test 3: 12 marks

The formula used by the head of maths to work the weighted average for all three tests combined was:

$$\text{Overall weighted average} = \frac{\text{Test1} \times 60}{100} + \frac{\text{Test2} \times 20}{60} + \text{Test3}$$

Using this formula we can work out the overall weighted score as:

$$\frac{55 \times 60}{100} + \frac{60 \times 20}{60} + 12$$

This works out to $33 + 20 + 12 = 65$ marks

Example 3:

A GCSE subject consists of a practical and theory exam. The practical exam has maximum raw marks of 20 and the final weighting given to this is 25%. The theory exam has a maximum raw mark of 80 and the final weighting given to this is 75%. If a pupil got 16 in his practical and 32 in his theory, what was the pupil's final weighted score?

Method:

For the practical exam: The pupil scored of 16 out of 20. This portion is equivalent to a weighted percentage of $16/20 \times 25\% = 20\%$

Similarly for the theory test the pupil scored 32 out of 80. This portion is equivalent to

$32/80 \times 75\% = 30\%$.

Hence the final score adding the two components together (20% +30%) gives us 50%

Formula

A formula describes the relationship between two or more variables. Consider a simple case first.

Example 1: The cost of claiming mileage to a parents evening at a school depends on how far the teacher lives, payable at 40p per mile and a fixed cost of £45 payable by this particular school for doing out of hours work.: We can write this as a formula,, $C = M \times 0.4 + 45$, where C represents the cost in pounds payable to the teacher by the school.

So for example if a teacher lives 15 miles from the school, he or she can claim 30 miles altogether for the journey to the school and back + £45 as shown below by the formula:

Using the formula we have $C = 0.4 \times 30 + 45 = 12 + 45 =$ £57

(Explanation of working out: Using BIDMAS we multiply before adding. So 0.4X30 =12, finally add 12 and 45 together to get 57)

Example 2:

(1) The formula for working out the distance depends on the speed and time taken in the appropriate units.

$D = S \times T$ where D is the distance, S the speed and T is the time.

What is the distance travelled if my speed is 60kmh and I travel for 1hour and 30 minutes.

1 hour 30 minutes corresponds to 1.5 hours so, using the formula, $D = 60 \times 1.5 = 90$ km

That is, the distance equals 90km

(2) The formula for working out the speed is given as Speed= Distance/Time

That is $S = D \div T$

Work out the average speed with which I travel, if I cover 100 miles in 2.5 hours.

Since $S = D \div T$, this means $S = 100 \div 2.5 = 40$ mph (Notice the units for the first example were in kilometers and units for the second example were in miles)

Example 3:

The formula for converting the temperature from Celsius to Fahrenheit is given by the formula: $F= \frac{9}{5}C +32$ (where C is the temperature in degrees Centigrade)

If the temperature is 10 degrees Celsius then what is the equivalent temperature in Fahrenheit?

Using the formula $F= \frac{9}{5}C +32$, and substituting 10 in place of C, we have $F= \frac{9}{5} \times 10 +32 = \frac{90}{5} + 32 =18+32 =50$. Hence, 10 degrees centigrade = 50 degrees Fahrenheit

Explanation of working out above: Remember we multiply and divide before adding and subtracting) There are no brackets to worry about. When working out $\frac{9}{5} \times 10 +32$, multiply 9 by 10 to get 90, divide this by 5 to get 18, finally add 18 and 32 together to get 50

Example 4: Convert 68 degrees Fahrenheit to degrees Celsius. The formula for converting the temperature from Fahrenheit to Celsius is given by:

$C= \frac{5}{9}(F-32)$, So to change 68 degrees Fahrenheit to degrees Celsius we can substitute for F in the formula $C=\frac{5}{9}(F-32)$, $C =\frac{5}{9}(68-32) = \frac{5}{9} x\ 36= 5 \times 4 =20$. Hence, 68 degrees Fahrenheit =20 degrees Celsius

Explanation of the working out above: Using BIDMAS we work out the bracket first. This gives us 68-32 =36. We now divide this by 9 and multiply by 5. Clearly 36÷9 =4 and finally 5×4 =20

We have seen that formulas can be important in conversion problems

Earlier we saw the formula: S = D ÷T, that is, Speed = $\frac{\textbf{Distance}}{\textbf{Time}}$.

Sometimes in the On screen questions you may be shown a distance time graph for a school coach trip and asked to work out average speed for a particular part of the journey and the time the coach was stationary. See example below.

Example: A school trip by coach to a heritage site leaves at 1200hrs from the school. The coach arrives at the destination at 1300hrs. It then stops so the pupils can look around the site. Finally after looking around the site it leaves and arrives back at school at 15:30hrs. (1) How long did the coach stop for? (2) What was the average speed on the return journey?

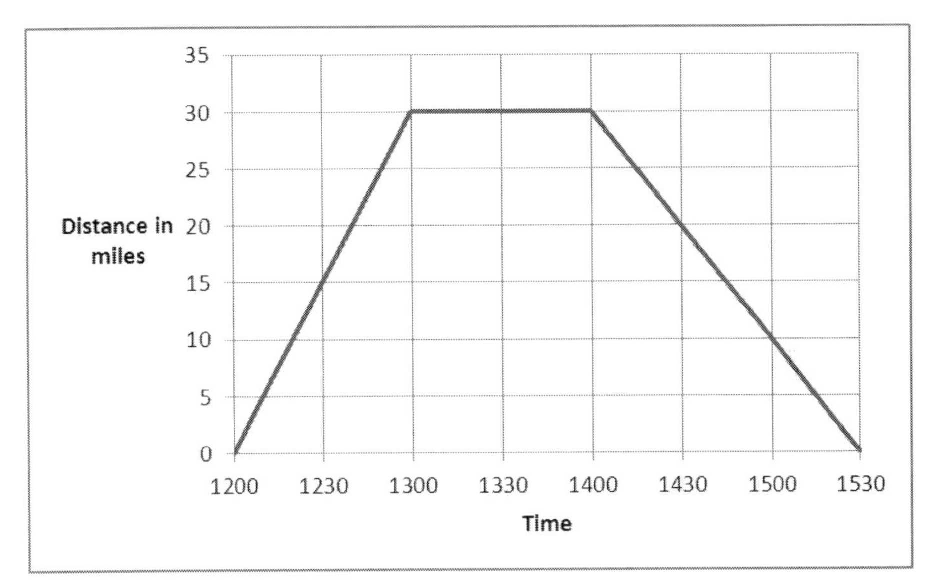

(1) From the distance-time graph above you can see it was stationary from 1300 – 1400hrs, which is 1hr

(Between these time intervals no further distance is covered, so it is stationary – see the vertical axis at 30 miles)

(2) The return journey starts at 1400hrs and ends at school at 1530hrs = 1.5 hrs.

Since $\textbf{\textit{Speed}} = \frac{\textit{Distance}}{\textit{Time}}$, this means speed = 30÷1.5 = 20 mph

You might find the following conversions useful to go through

(Typically, the QTS Numerical test questions give you the conversion formula in the relevant questions)

1 km = 5/8 mile

1 mile =8/5 km

1kg =2.2 lb (approximately)

1 gallon =4.5 litres (approximately)

1 inch = 2.54 cm (approximately)

More Formulas

Although you don't need to know how to work out Areas, Perimeters and Volumes for the QTS exam it is useful to know the formulas for these common shapes as it also gives you more practice in using formulas.

(1) Rectangle

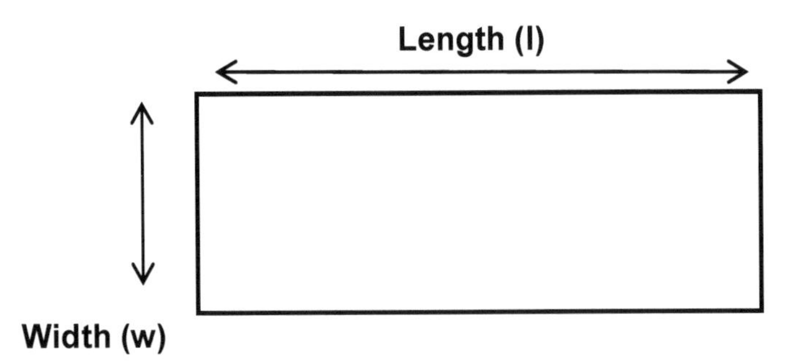

Area of a rectangle = Length × Width or l × w

Perimeter of a rectangle = 2l + 2w (distance around the rectangle)

Note: Area is measured in units squared, e.g. cm^2 or m^2 and perimeter (distance all round a shape) is measured in the appropriate units e.g. cm or m

Mental Arithmetic question based on areas

Example 1: Find the area and perimeter of a rectangle whose length is 12 cm and width is 5cm

Method: Area of a rectangle = $l \times w = 12 \times 5 = 60\ cm^2$.

Perimeter =$2l + 2w = 2\times12 + 2\times5 = 24 + 10 = 34$ cm

Note: For a square the length and width are obviously the same so the formula simplifies to l×l for the area and 4l for the perimeter, where l is the length of each side of the square

Example 2: Find the area of a square whose sides are 1m. Give your answer in cm^2. Area of square in metres squared = $1\times1 = 1\ m^2$, but 1 m = 100cm. So area in cm^2 = 100X100 =10000 cm^2

(2) Triangle

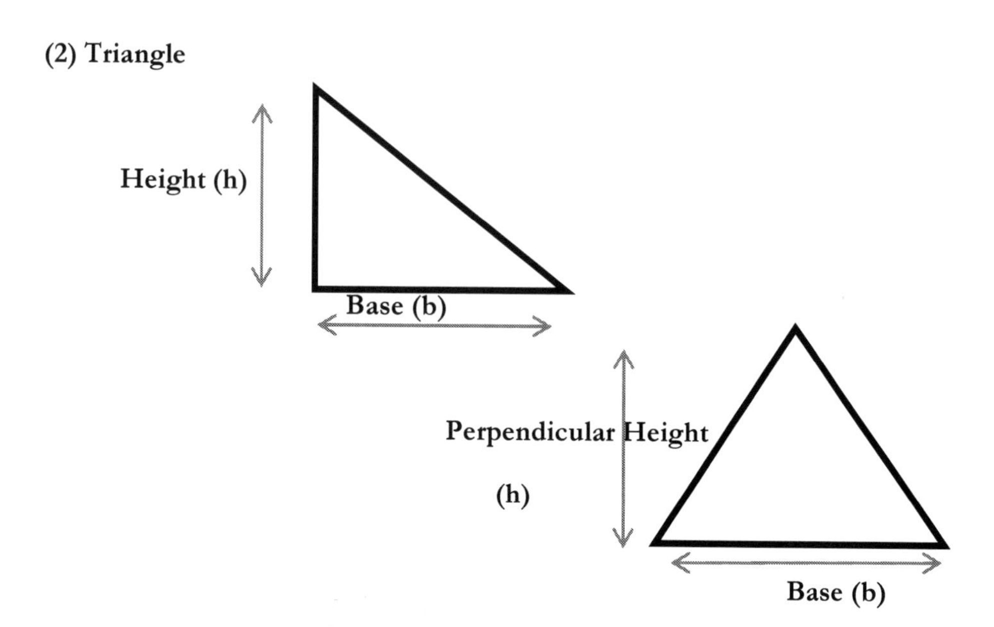

Area of a triangle =1/2 × base × height or 1/2×b×h (The height is the perpendicular height relative to the base)

> **Example**: Find the area of a triangle whose base is 5m and height is 8m
>
> Method: Area of a triangle = 1/2 × base × height
>
> Substituting the values for base and height we get Area = 1/2 × 5 × 8 = 1/2 × 40 =20 m^2

Area of a circle is πr^2 (this means the value of π(pi) multiplied by radius squared)

Circumference of a circle (distance all the way round a circle) = 2πr or πd

2πr (2 x π(pi)) x radius or π x diameter

Note: Diameter of a circle = 2 × Radius

Approximate value of π =3.14

Volume of a cuboid (or a box)

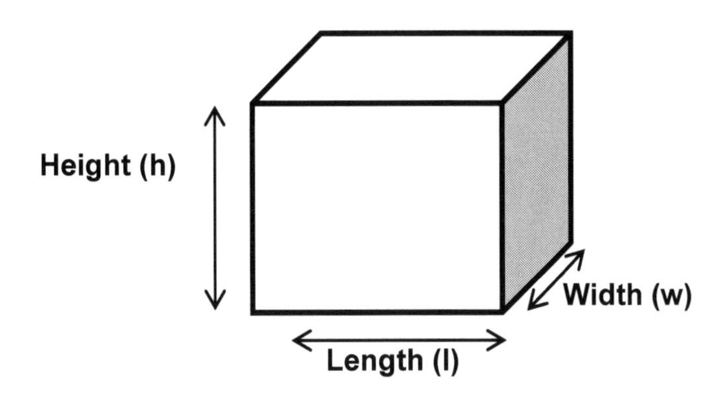

Volume of a cuboid is Height × Length × Width or V = h×l×w (units cubed e.g. cm^3 or m^3, etc)

Example: Find volume of a box whose height =3m, length =5m and width = 6m. **Method:** Volume of a cuboid (box) = h×l×w = 3×5×6 =15 × 6 = 90 m^3

Mock Mental Arithmetic Tests

Mental Arithmetic - Mock Test 1

Question 1

There are 21 pupils in a class. 3 pupils go for extra Numeracy lessons in a given lesson. What is the fraction of the pupils that remain in class? Give your answer in its lowest terms.

Answer:

Question 2

2500 millilitres of liquid is divided into 20 containers. How many millilitres of liquid does each container have?

Answer: ml

Question 3

A school trip in France involves walking 24 Km every day.

If 8km is approximately equal to 5 miles, estimate how many miles the daily walk consist of?

Answer:  Miles

Question 4:

A primary school has 80 pupils in year 6. Twenty pupils have a reading age that exceed their actual age. What is the percentage of pupils who do not exceed their reading age?

Answer: %

Question 5:

18 pupils are asked to collect £3.50 each for a charity. All of them succeed.

What is the total amount collected?

Answer: £

Question 6

A group activity consists of 16 tasks. Each task lasts 15 minutes. How many hours will this group activity last?

Answer: Hours

Question 7

A maths lesson begins at 10:50. The teacher introduces the topic for 6 minutes, there is a warm up exercise for 18 minutes and finally work is done on the new topic for the last 26 minutes. When does the lesson end, give your answer using the 24-hour clock?

Question 8:

A school calculated that it had given merits to boys and girls in the ratio of 1:3. There were a total of 680 merits given. How many merits did the boys get?

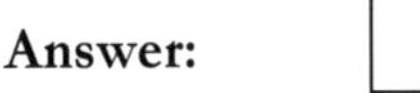

merits

Question 9

In a science exam 65% of the pupils achieved a level 5 in Key Stage 2. There were 140 pupils in this year group. How many pupils did not achieve level 5?

Answer: ☐ pupils

Question 10

A teacher has to see 16 parents for 12 minutes each to discuss pupil progress. In addition, there is a 25-minute break. How long does the parents' session last in hours and minutes?

Hours Minutes

Answer:

Question 11

A teacher attends a course which is 80 miles away. She is allowed to claim travel expenses for the journey there and back at 40p per mile. How much is the teacher allowed to claim?

Answer: £ ☐

Question 12

All the year 9 pupils are taken on a day trip to the local museum. There are 48 pupils altogether. There is requirement for at least one adult per 8 pupils. What is the total number of people on this trip?

Answer: ☐

Mental Arithmetic - Mock Test 2

Question 1

In a Junior School there are 440 pupils and 35% have free school dinners. Work out the number of children who do not have free school dinners.

Answer: ☐

Question 2

240 pupils sat a GCSE maths exam. The fraction of pupils who get Grade C or above is $\frac{3}{5}$. How many pupils get Grade D or below?

pupils

Question 3

28 pupils go to visit the Science Museum in London. Each pupil has to make a contribution of £3.25 which the teacher collects. How much is the total contribution that is collected?

Answer: £ ☐

Question 4:

In a sixth form college of 340 pupils, each pupil has to have a planner. 25% of the pupils get their planner free. The remaining pupils pay £2.00 each for their planner. How much in total do the students spend on planners?

Answer: £ ☐

Question 5:

A double lesson ends at 1425. If each lesson was 45 minutes, when did the lesson start?

Answer: [ ]

Question 6

A pupil scores 45.5% in Test1 and 64.5% in Test2. What was the pupil's average mark, assuming they were weighted equally?

Answer: [ ] Marks

Question 7

The ratio of girls to boys in a school is 4:5. Assuming there are 900 pupils altogether, how many boys are there?

Answer: [ ] Boys

Question 8:

A practical session in Science consists of 3 very short experiments. The first experiment takes 56 seconds, the second 47 seconds and the third one takes 39 seconds. What is the total time taken in minutes and seconds for all the three experiments?

	Minutes	Seconds
Answer		

Question 9

In a maths exam 45% of the pupils achieved GCSE at Grade C or above. There are 180 pupils in this year group. How many pupils achieved at least a Grade C?

Answer: [ ] pupils

Question 10

There are 11 girls and 14 boys in a class. What is the percentage of boys in this class?

Answer: [ ] %

Question 11

The head of English orders 80 books at £2.99 each from a remaining budget of £780.50. How much does the teacher have left after ordering the books?

Answer: £ [ ]

Question 12

One gallon is approximately 4.5 litres. How many gallons are there in 54 litres?

Answer: [ ] Gallons

Mental Arithmetic - Mock Test 3

Question 1

What is 452 divided by 0.1?

Answer: [ ]

Question 2

A school trip to France and back covers 640 kilometers. Assuming 5 miles equals 8 kilometers, how many miles in total is the trip?

Answer: [ ] Miles

Question 3

There are 780 pupils in a school. If the proportion of pupils who have free school dinners is 15%. How many pupils are not eligible for free school dinners?

Answer: [ ] Pupils

Question 4:

A primary school has 80 pupils in year 5. Ten pupils have a reading age that is below their actual age. What is the percentage of pupils who have a reading age that is below their actual age? Give your answer as a decimal.

Answer: [ ] %

Question 5:

A teacher has planned to see 16 parents to discuss their children's progress this term. Each parent's session lasts 15 minutes. In addition there is a break of 20 minutes. If the parents evening starts at 1730 when does it end?

Answer:

Question 6

150 pupils took the GCSE English exam. The percentage of pupils who got grade C or better in 2007 was 60% and the percentage of pupils who got a grade C or better in 2008 was 62%. How many more pupils passed in 2008 compared to 2007?

Answer: Pupils

Question 7

Two geography classes get together to watch a video. The first class has 14 boys and 12 girls. The second class has 14 girls and 10 boys. When the class is combined what is the percentage of girls as a proportion of the total number of pupils?

Answer: %

Question 8

Pupils in a school are asked to contribute £1.25 each to an educational charity. 340 pupils actually contributed. How much is collected?

Answer: £ 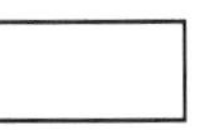

Question 9

In a History class $\frac{2}{5}$ of the pupils achieved a level 5 in Key Stage 2. In another class $\frac{1}{4}$ of the pupils achieve this level. What is the total fraction for both classes combined that achieves this level?

Answer:

Question 10

A pupil's scored 35%, 47% and 44% respectively in a three different maths tests. What was the pupil's mean mark?

Answer:  Marks

Question 11

A teacher attends a seminar which is 27 miles away from her home. She is allowed to claim travel expenses for the journey there and back at 40p per mile. How much is the teacher allowed to claim?

Answer: £

Question 12

5 teachers organize an outing for 35 pupils. What is the percentage of teachers in the whole group?

Answer: %

Mental Arithmetic - Mock Test 4

Question 1

School extra curricula activities are planned for 5 classes of 23 pupils and 4 classes of 26 pupils. How many pupils in total are involved?

Answer: [ ] Pupils

Question 2

A student scores 45 marks in a standardized test. After some weeks of additional tuition his scores improve by 30%. What marks does he now get in a similar test?

Answer: [ ] Marks

Question 3

In a science experiment a teacher needs exactly 85 cubic centimeters of blue coloured water for an experiment. The teacher has prepared a large container which has 1700 cubic centimeters of blue coloured water. How many pupils can do this experiment in this lesson?

Answer: [ ] Pupils

Question 4:

A school day ends at 1520. In the afternoon there are only 2 sessions of 45 minutes each. When does the afternoon session start?

Answer: [ ] s

Question 5:

What is 12.5% of 360 Kilograms?

Answer: [ ] **Kg**

Question 6

Eight out of 25 pupils scored 75% or more in a test. What percentage of pupils scored less than 75%?

Answer: [ ] %

Question 7

What is 565 divided by 0.5?

Answer: [ ]

Question 8:

In a class of 28 pupils, 4 pupils require additional help in numeracy. What is the fraction of pupils that do not require help in Numeracy? Give your answer in its lowest terms.

Answer: [ ]

Question 9

In an English test a pupil achieved 27 out of 45 marks. What was the percentage mark that the pupil received in this test?

Answer: %

Question 10

A school trip to an art gallery requires two contributions. One bit for the coach trip and other for the entrance fee. The total contribution required was £12.50. If one fifth of the cost was the entrance fee, how much was the contribution for the coach trip?

Question 11

The head of maths at a school implements a homework policy of 1.5 hours per week in total. The homework is to be done during the weekdays in the evenings. What is the mean time in minutes spent doing homework per weekday?

Answer:  Minutes

Question 12

A history class joins a Geography class for a school trip. The total number of pupils in this trip is 56. If $\frac{3}{7}$ of the pupils were from the Geography class, how many pupils were there from the History class?

Answer: Pupils

Mental Arithmetic - Mock Test 5

Question 1

In a PE session token rewards are awarded to pupils who come first, second and third in a 200 Metre run. Altogether there are 12 tokens that are awarded in the ratio of 1:2:3, the highest number of tokens being awarded to the pupil who comes first, the second highest to the pupil who comes second and finally the lowest number of tokens to the pupil who comes third. How many tokens does the pupil who comes second get?

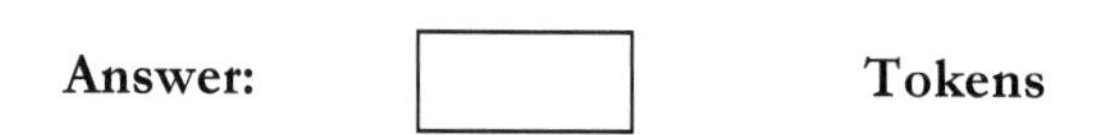

Question 2

21 pupils from a year 10 group join 19 pupils from a year 9 group to go on a school outing with 8 teachers. What is the fraction of teachers compared to the whole group that go on this outing?

Answer:

Question 3

Work out 12.5% of 2 Kilograms. Give your answer in grams.

Answer: Grams

Question 4

There are 27 pupils in a class. 4 out of 9 pupils are girls. How many boys are there?

Answer: 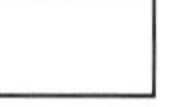 Boys

Question 5

There is a rectangular lawn in the school grounds. The length is 8.5m long and the width is 4m. What is the area of the rectangular lawn?

Answer: Metres Squared

Question 6

What is 52.2 multiplied by 1000?

Answer:

Question 7

A coach can accommodate 52 people. There are 42 pupils who go on a school outing in this coach. During the outing there has to be is one teacher for every 6 pupils. How many vacant seats are there?

Answer:  Vacant Seats

Question 8

In a year 6 numeracy test, 27 out of 45 pupils obtain a level 5. What is the percentage of pupils that do not achieve level 5?

Answer: %

Question 9

A teacher plans to show a video in a double lesson. The teacher introduces the topic for 7 minutes, the video itself lasts for 32 minutes and the discussion at the end of the video lasts 19 minutes. If the double lesson starts at 13:40 and ends at 14:50, how many minutes are left before the end of the lesson after the discussion ends?

Answer: Minutes

Question 10

A teacher needs to write reports in her subject for 47 pupils. Each report takes an average of 11 minutes to write. She also spends an average of 3 minutes checking each report. How long in hours and minutes does it take the teacher to finish the report writing tasks?

	Hours	Minutes
Answer		

Question 11

A school has 950 pupils. 15% play netball, 35% play football and the rest play hockey. How many pupils play hockey?

Answer: ☐ Pupils

Question 12

150 pupils get GCSE grade C or above in 2005, a further 30 pupils get GCSE grade C or above in 2006. What is the percentage increase of students who perform at this level from 2005 to 2006?

Answer: ☐ %

Answers to Mock Mental Arithmetic Test 1

Question 1

Answer: $\frac{6}{7}$

Method: Number of pupils that remain in class =18, corresponding fraction =$\frac{18}{21}$ = $\frac{6}{7}$ (divide both numerator, 18 and denominator, 21 by 3)

Question 2

Answer: 125 ml

Method: Number of millilitres in each container =2500÷20 =125

Question 3

Answer: 15 miles

Method: If 8km = 5miles, then 24 km = $\frac{5}{8}$ × 24 = 5 × 3 =15 miles

(Notice the denominator 8 goes into 24 three times)

Question 4

Answer: 75%

Method: 20 pupils out of 80 exceed their reading age. This means 60 do not.

Percentage = $\frac{60}{80}$ x 100 = $\frac{3}{4}$ x 100 =75

(Notice $\frac{60}{80}$ simplifies to $\frac{3}{4}$, by dividing both numerator and denominator in the original fraction by 20)

Question 5

Answer: £63

Method: 18 × £3.50

= £18×3 + 18 × 50p

= £54 + £9 = £63

Question 6

Answer: 4 hours

Method: Total number of minutes = 16 × 15

Total hours = (16 × 15) ÷ 60 = $\frac{16X15}{60} = \frac{16}{4} = 4$

(Divide top and bottom in the fraction: $\frac{16X15}{60}$ by 15)

Question 7

Answer: 11: 40

Method: Total time in lesson = 6 + 18 + 26 = 24 + 26 = 50 minutes

10:50 plus 50 minutes = 11:40

Question 8

Answer: 170

Method: In the ratio 1:3, the total number of parts are 4 (add 1 and 3)

This means boys got $\frac{1}{4}$ of the merits. Hence $\frac{1}{4}$ of 680 = 170

(You can halve 680 twice. Half of 680 =340, then half of 340 =170)

Question 9

Answer: 49

Method: The percentage of pupils who do not achieve level 5 is 35%

35% of 140 pupils = (30% + 5%) of 140 pupils

10% of 140 =14

30% =3 X 14 =42

5% is half of 10% = 7 pupils

Hence 35% = 42 + 7 = 49 pupils

Question: 10

Answer: 3 hours 37 minutes

Method: Total time taken = 16×12 + 25

=16×12 =16×10 + 16×2 =160 +32 =192

Now we need to add 25 to 192 to get 217 minutes

217 minutes = 180 +37 = 3hours and 37 minutes

Question 11

Answer: 64

Method: Total distance allowed = 160 miles

Expenses allowed = 160×40p =£160×0.4 = £16×4 =£64

Question 12

Answer: 54

Method: There are 48 pupils altogether. 1 adult per 8 pupils means 6 adults are needed. (Since 48 ÷ 8 =6) This means there are a total of 48 + 6 people = 54 people on this trip

Answers to Mock Mental Arithmetic Test 2

Question 1

Answer: 286

Method: Number of children who do not have school dinners = 65%

10% of 440 =44

60% = 6×44 = 6×40 + 6×4 =240+24 =264

5% = 22

So 65% = 264 + 22 = 286

Question 2

Answer: 96

Method: Pupils who get Grade D or below $=\frac{2}{5}$ of 240 $=\frac{2}{5} \times 240 = \frac{480}{5} = 96$

(To divide by 5 quickly, you can double 480 and divide by 10, that is $\frac{960}{10}$

=96

Question 3

Answer: 91

Method: Total contribution collected = £3.25 × 28 = £3×28 + 28 × 25p= £84 +£7 = £91

(Since 25p = a quarter of £1, then 28×25p = £7)

Question 4

Answer:

£510

Method: 75% of the pupils pay for their planners. 75% of 340 = 50% of 340 + 25% of 340

= 170 + 85 = 255 pupils

Since each pupil pays £2.50, then the total cost = £255 × 2 = £510

Question 5

Answer: 12:55

Method: The double lesson lasts 90 minutes = 1hour 30 minutes

Lesson starts at 14:25 – 1hour 30minutes = 12:55

Question 6

Answer: 55%

Method: Total marks obtained = 45.5 + 64.5 =110

Mean = 110 ÷ 2 = 55%

Question 7

Answer: 500 boys

Method: Ratio of girls to boys is 4:5 which means there are a total of 9 parts. Each part is one ninth or 900÷9 =100

Since there are 900 pupils altogether, the number of boys = 5 × 100 =500

Question 8

Answer: 2 minutes 22 seconds

Method: Total time taken for the 3 experiments = 56 + 47 + 39 = 142 seconds

142 seconds = 120 + 22 = 2 minutes and 22 seconds

Question 9

Answer: 81 pupils

Method: 45% of 180 pupils = (40% + 5%) of 180

Since 10% =18, then 40% = 4×18 =72

5% = 9, so 45% = 72 +9 =81 pupils

Question 10

Answer: 56%

Method: Total number of pupils = 25.

Percentage of boys = $\frac{14}{25} \times 100 = 14 \times 4 = 56\%$

Question 11

Answer: £541.30

Method: Cost of books = £80×3 – 80p = £240 – 80p = £239.20

Budget left = £780.50 – 239.20 = £541.30

Question 12

Answer: 12

Method: Number of gallons in 54 liters = 54÷4.5 =540÷45 = 108÷9 = 12

(Notice when dividing 54 by 4.5, to make it easier to simplify multiply both 54 and 4.5 by 10 to give 540÷45, then simplify further by dividing both 540 and 45 by 5 to give 108÷9, finally this gives you 12)

Answers to Mock Mental Arithmetic Test 3

Question 1

Answer: 4520

Method: 452÷0.1 means 452 divided by one-tenth. The answer should be 10X bigger,

So 452÷0.1 = 4520

Question 2

Answer: 400 miles

Method: 640 km = $\frac{5}{8}$ × 640 miles = 5 × 80 = 400 miles

Question 3

Answer: 663

Method: Pupils not eligible for free school meals = 85%

10% of 780 = 78, 80% = 78×8 = 70×8 + 8×8 = 560 +64 =624

Since 10% =78, 5% = 39, Finally 85% = 80% + 5% = 624 + 39 = 663

Question 4

Answer: 12.5%

Method: Fraction of pupils who have their reading age below their actual age is 10 out of 80.

Since, $\frac{10}{80} = \frac{1}{8}$

Equivalent percentage = $\frac{1}{8}$ × 100% = $\frac{100}{8} = \frac{25}{2}$ = 12.5%

Question 5

Answer: 21:50

Method: Total time taken = 16×15 minutes = $16 \times 10 + 16 \times 5 = 160 + 80 = 240$ minutes plus 20-minute break

$240 \div 60 = 4$ hours plus 20-minute break. Since parent's evening starts at 17:30 and lasts 4hours 20 minutes, it ends at 21:50

Question 6

Answer: 3

Method: 2% more pupils from 2007 to 2008.

2% of 150 = $\frac{2}{100}$ x 150 = $\frac{300}{100} = 3$

Question 7

Answer: 52%

Method: Total number of pupils = 50, total number of girls = 12 + 14 =26

Percentage of girls = $\frac{26}{50}$ x 100 = 26 x 2 = 52%

Question 8

Answer: £425

Method: Total contribution = 340 × £1.25 = £340 + 25pX340 = £(340 + $\frac{340}{4}$)= £340 +£85 = £425

Question 9

Answer: $\frac{13}{20}$

Method: Add the two relevant fractions, $\frac{2}{5} + \frac{1}{4} = \frac{8+5}{20} = \frac{13}{20}$

Question 10

Answer: 42%

Method: Mean mark = (35 + 47 + 44)/ 3 = 126/3 = 42

Question 11

Answer: £21.60

Method: Total mileage allowed = 27×2 = 54 miles

Can claim 54X40p = £54X0.4 = £21.60

Question 12

Answer: 12.5%

Method: Total people on outing = 40

Percentage of teachers = $\frac{5}{40}$ × 100 = 500/40 = 50/4 = 25/2 =12.5%

Answers to Mock Mental Arithmetic Test 4

Question 1

Answer: 219

Method: Total pupils = $23\times5 + 26\times4 = (20 + 3)\times5 + (20 + 6)\times4$

$= 20\times5 + 3\times5 + 20\times4 + 6\times4 = 100 + 15 + 80 + 24 = 180 + 15 + 24 = 195 + 24$ $= 219$

Question 2

Answer: 58.5%

Method: 10% of 45 = 4.5 marks, so 30% = $3\times4.5 = 13.5$

New mark = 45 +13.5 = 58.5%

Question 3

Answer: 20

Method: Number of pupils who can do this experiment = $1700\div85$, double 1700 & 85 to make the calculation simpler if you prefer. This is the same as $3400\div170 = 340\div17 = 20$

Question 4

Answer: 13:50

Method: 2 sessions of 45 minutes = 90 minutes = 1 hour 30 minutes

So lesson starts at 15:20 minus 1 hour and 30 minutes = 13:50

Question 5

Answer: 45

Method: 50% =180kg, 25% =90Kg, 12.5% =45Kg

Question 6

Answer: 68%

Method: If 8 out of 25 scored 75% or more, then 17 out of 25 scored less than 75%. Now $\frac{1}{25}$ ×100= 4%, this means $\frac{17}{25}$ x 100 = 17 × 4 = 68%

Question 7

Answer: 1130

Method: 565 ÷0.5 (this means how many halves are there in 565)

So, 565 ÷0.5 = 565×2 = 1130

Question 8

Answer: $\frac{6}{7}$

Method: Number of pupils that do not require help is 24. Hence the fraction that does not require help is $\frac{24}{28}$, now simplify this fraction by dividing numerator and denominator by 4 to get $\frac{6}{7}$

Question 9

Answer: 60%

Method: % mark = $\frac{27}{45}$ × 100, consider $\frac{27}{45}$ to simplify this divide both numerator and denominator by 9. This reduces $\frac{27}{45}$ to $\frac{3}{5}$ Finally, $\frac{3}{5}$ × 100 =60%

Question 10

Answer: £10.00

Method: $\frac{1}{5}$ of £12.50 = $\frac{1}{5}$ of £(10 +2.50) = £2 + 50p = £2.50

So the coach trip costs £12.50 - £2.50 = £10.00

Question 11

Answer: 18 minutes

Method: 1.5 hours = 1 hour 30 minutes = 90 minutes

Mean time spent per day =90/5 =18 minutes

Question 12

Answer: = 32

Method: If $\frac{3}{7}$ were from the Geography class then $\frac{4}{7}$ were from the History class.

The number of pupils who were from the history class was $\frac{4}{7}$ of 56

Now $\frac{1}{7}$ of 56 = 8, so = $\frac{4}{7} \times 56 = 4 \times 8 = 32$

Answers to Mock Mental Arithmetic Test 5

Question 1

Answer: 4

Method: Total parts in the ratio 1:2:3 = 6

Since there are 12 tokens awarded, each part is worth 2 tokens

The pupil who comes second has 2 parts, which means this pupil gets 4 tokens

Question 2

Answer: $\frac{1}{6}$

Method: Total persons on this outing =21 +19 +8 = 48. Since there are 8 teachers the fraction of teachers compared to the whole group =$\frac{8}{48}$,

which simplifies to $\frac{1}{6}$

Question 3

Answer: 250 grams

Method: 2kg = 2000 grams, 50% =1000, 25% =500 and 12.5% =250 grams

Question 4

Answer: 15

Method: $\frac{4}{9}$ are girls, which means $\frac{5}{9}$ are boys. Since $\frac{1}{9}$ of 27 = 3, then $\frac{5}{9}$ of

27 = 5×3 =15

Question 5

Answer: 34 Metres squared

Method: Area = length × width = 8.5 × 4 = (8 +0.5) × 4 = 8×4 + 0.5×4 =32 +2 =34

Question 6

Answer: 52200

Method: 52.2 × 1000 = 52200

When 52.2 is multiplied by 1000, the answer is 1000 times bigger. An easy way to do this is to move the decimal place 3 places to the right which means 52.2 × 1000 = 52200.0 or 52200)

Question 7

Answer: 3 seats

Method: There are 42 pupils. If there is one teacher for every 6 pupils, then it means there are 7 teachers (6×7 =42). This means altogether 42 + 7 =49 people travel on the coach. As the coach can accommodate 52 people, there are 3 vacant seats remaining.

Question 8

Answer: 40%

Method: Number of pupils who do not obtain level 5 is 18. So, the percentage that did not achieve level 5 is $\frac{18}{45}$ ×100. By dividing top and

bottom of the first bit by 9, this simplifies to $\frac{2}{5}$ × 100 = 200/5 = 40%

Question 9

Answer: 12

Method: Total session lasts 7 + 32 + 19 = 58 minutes

13:40 + 58 minutes = 14:00 + 38 minutes = 14:38

Since the lesson ends at 14:50 and session ends at 14:38, there are 12 minutes left

Question 10

Answer: 10 hours 58 minutes

Method: Total time taken to write & check reports = 47×11 + 47×3 = 517 +141 = 658 minutes

Now 658 minutes = 600 + 58 = 600÷60 hours+ 58 minutes

=10hours and 58 minutes

Question 11

Answer: 475

Method: Percentage who play netball and football = 15% + 35% =50%. This means 50% play hockey. As there are 950 pupils in total, 50% of 950 = 475 pupils

Question 12

Answer: 20%

Method: Percentage increase is $\frac{30}{150} \times 100 = \frac{1}{5} \times 100 = 20\%$

Chapter 6 Statistics

Mean, Median, Mode and Range

First consider the different types of 'averages'.

That is Mean, Median, Mode and Range (You can try to remember these as: MMMR)

Mean: The sum of the numbers in a data set divided by the number of values in the
Set

Median: The middle number of a data set when listed in order .

Mode: The most frequently occurring number or numbers in a data set

Range: The difference between the highest and the smallest numbers in a data set

Example 1:
Find the mean value of the following data set:
2, 7, 1, 1, 7, 8, 9

Method: Find the sum first
$2 + 7 + 1 + 1 + 7 + 8 + 9 = 35$
Now divide this total by 7, since this is the total number of numbers
So, $35/7 = 5$
Hence, the mean value of this data set is 5

Example 2:
Find the median of 3, 7, 1, 8, and 6

Method: First re-order from smallest to biggest, re-writing the numbers we have: 1, 3, 6, 7, 8
Clearly the middle number is 6.
Hence, the median is 6

Example 3:
Find the median of 3, 6, 7, 1, 8 and 5

Method
First re-arrange to get 1, 3, 5, 6, 7, 8
Notice, in this case the middle number is between 5 & 6
So the median is (5 + 6)/2 = 5.5

Example 4:
Find the Range of the data set 3, 5, 7, 1, 8, and 11

Method: Find the difference between the biggest and smallest numbers
So the Range = 11 – 1 = 10

Example 5:
Find the Mode of the following numbers:
1, 4, 4, 4, 7, 8, 9, 9, 11, 12

Method: Find the most frequently occurring number. The most frequently occurring number is 4.
Hence the Mode is 4

Example 6:
Find the mode of 1, 3, 3, 3, 3 5, 5, 5, 5, 8, 8, 9

Method: As before find the most frequently occurring number(s)
Clearly there are two modes here. Both '3' and '5' occur most frequently, the same number of times, so we say this is a bi-modal distribution. That is, a distribution with two modes, namely 3 and 5.

Typical On screen questions (although these examples below may have two parts to show different aspects of the question)

Example 1:

The set of data below is the result in a class maths test showing the marks out of 10 for a group of 27 pupils. The teacher wants to find (1) the mode and (2) the mean mark for this test.

Maths marks	No of pupils (frequency)	No. of pupils × Maths marks (frequency × marks)	
9	0	$0 \times 9 = 0$	
8	1	$1 \times 8 = 8$	
7	2	$2 \times 7 = 14$	
6	5	$5 \times 6 = 30$	
5	8	$8 \times 5 = 40$	
4	7	$7 \times 4 = 28$	
3	4	$4 \times 3 = 12$	
2	0	$0 \times 2 = 0$	
1	0	$0 \times 1 = 0$	
Totals	27	0+8+14+30+40+28+12+0+0=132	

(1) The mode is simply the most frequently occurring mark. In this case it is 5 marks

(Since 8 pupils get this, clearly it is the most common result)

(2)To work out the mean we need to work out the sum of all (pupils × marks) and divide it by the total number of pupils as shown below:

Sum all (pupils X marks) as shown above =132

The total number of pupils who took the test is 27

The mean mark in this test was thus $132 \div 27 = 4.9$ (to 1 decimal place)

Example 2:

The set of data below shows GCSE grade results in a particular subject for a group of 25 pupils in 2010 in a certain school.

B A C A A*

C D D G C

B C G E C

F G A C B

C B D B C

In addition, point scores for each grade are as shown

A*	A	B	C	D	E	F	G
58	52	46	40	34	28	22	16

The Head teacher asked the responsible teacher to find out what the mean point score was and to compare it to the mode.

Method

See working out in the table and below

Step1: Work out the frequencies (the number of pupils) getting a particular grade, simply add up the number of times that particular grade occurs in the set of data given **above**. So for example there was only 1 occurrence of A*, there were 3 occurrences of Grade A, 5 occurrences of Grade B and so on.

Grades	A*	A	B	C	D	E	F	G
Point Score	58	52	46	40	34	28	22	16
Frequency (Number of students achieving this grade)	1	3	5	8	3	1	1	3
Frequency X Point Score	58×1 = 58	52×3 =156	46×5 =230	40×8 =320	34×3 =102	28×1 =28	22×1 =22	16×3 =48

Step2: Work out Frequency × point score for each grade as shown in the last row.

Step3: Add up all the Frequency × point score values, 58 +156 +230 +320 +102 +28 +22 +48 = 964

Step4: To work out the mean simply divide 964 by the total frequency (total number of pupils)

So 964÷25 = 38.6

The mean point score is approximately 39 (rounded to the nearest unit). This value shows that the average grade by this measure is approaching C.

However, the mode (most frequently occurring grade) is in fact a grade C, since 8 pupils got this.

You can see that if the school represented its data as a modal value (mode), then the grade appears slightly better than if it uses the mean!

Trends

Trend questions typically involve identifying whether there is an increase or decrease in values such as percentages, grades etc. over a period of time. **See line graph later for an example**. Here we will mostly look at cases of sequences that increase or decrease either arithmetically (that is by the same amount) or geometrically that is for example doubling, or halving or increasing by some power law every time. **It is very probable that you will get questions that involve arithmetic sequences only, typically concerning reading ages.**

Before we consider a typical QTS question see if you can follow the sequences below and identify the patterns:

(1) 4, 7, 10, 13, __ , __ (you probably noticed this involves adding 3 to each number to find the next number, the missing numbers are 16 and 19)

(2) 21, 17, 13, 9, 5, __ , __ (in this case subtract 4 from each number to find the next number, the missing numbers are 1 and -3)

(3) 36, 18, 9, __ , __ (halve each number to find the next number, the missing numbers are 4.5 and 2.25)

(4) -6, -4, -2, 0, __ , __ (each number increases by 2, so the missing numbers are 2 and 4)

(5) 1, 4, 9, 16, 25, __, __ (square each integer (whole number), 1×1, 2×2, 3×3, etc so the missing numbers are 6×6 =36, 7×7 =49)

(Notice this last sequence is neither arithmetic nor geometric. It is simply the square of natural numbers 1, 2, 3, 4, 5, etc.)

Example of a typical question involving trends

A teacher in a primary school records the reading age for a group of 7 pupils over a five month period. In each month the recording done is that of the reading age minus the actual age for each pupil. Point and click (or in this case put a X in the last column) of the pupil(s) that show a consistent trend of improvement every month in the difference between reading age and their actual age over a five-month period.

	Reading Age minus actual age (months)					
Pupil	**Month 1**	**Month 2**	**Month 3**	**Month 4**	**Month 5**	**Select Pupil by putting 'X' in the appropriate row**
A	**2**	**3**	**5**	**5**	**6**	
B	**2**	**2**	**3**	**3**	**4**	
C	**-5**	**-4**	**-3**	**-3**	**-3**	
D	**1**	**2**	**2**	**3**	**4**	
E	**-5**	**-3**	**-1**	**1**	**3**	
F	**5**	**5**	**6**	**6**	**6**	
G	**-8**	**-7**	**-5**	**-5**	**-4**	

You can see from the table above that pupil E is the only pupil that shows a **consistent** increase in reading age **every** single month. In this case it is 2 months. (From -5 to -3, from -3 to -1, from -1 to 1 and finally from 1 to 3)

Notice: One could argue that there is a general improvement for every single pupil over the time period considered. However, since the question asks specifically for a consistent trend improvement every single month then only pupil E satisfies these criteria.

Pie Charts

When data is represented in a circle this is called a pie chart. Basically you need to remember that a full circle or 360 degrees represents all the data (or 100% of the data). Half a circle or 180 degrees represents half the data (or 50% of the data), and similarly 25% of the data is represented by 90 degrees or a quarter of a circle. Essentially, each sector or slice of the pie chart shows the proportion of the total data in that category.

Example 1:

The pie chart below shows the percentage of pupils who got different grades in GCSE maths in a given year at a particular school. If 28 pupils took GCSE Maths, How many pupils got a grade B or better?

Method:

As illustrated the GCSE results in Maths in a particular year show that 25% got grade B or above.

(Since a quarter of a circle corresponds to 25%)

This means a quarter of the 28 students attained this which corresponds to 7 pupils.

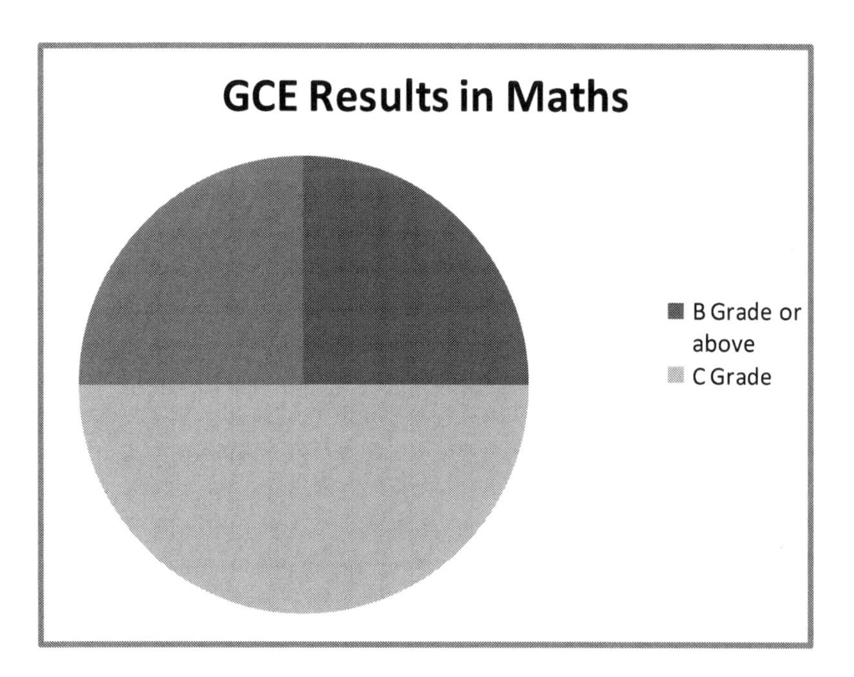

Example 2:

The destination of 120 pupils who leave year 11 in School B in 2012 is represented in the pie chart below. The numbers outside the sectors represent the number of pupils

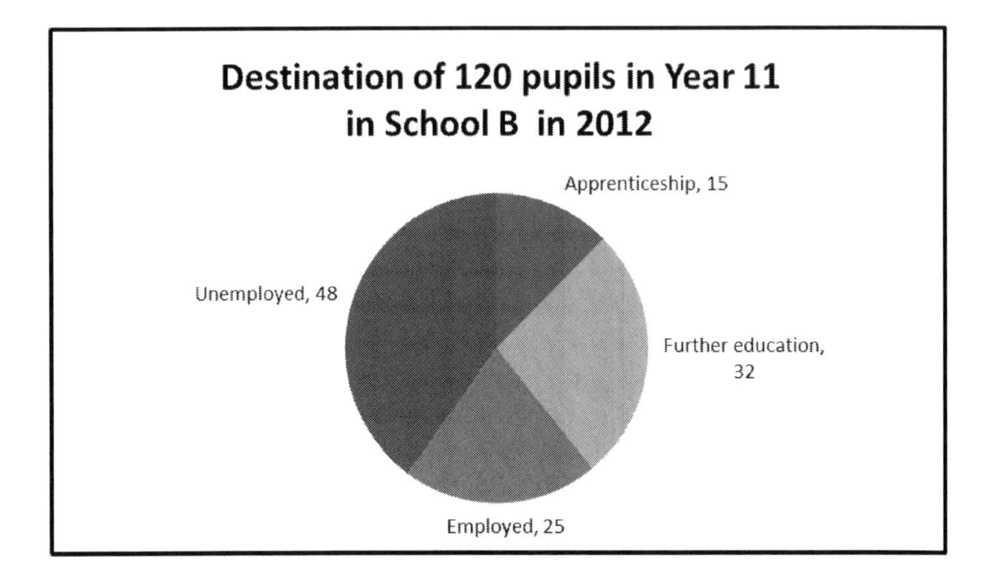

(1) What is the percentage of pupils who are unemployed?

Method:

The number of pupils out of 120 that are unemployed is 48. So the percentage of pupils who are unemployed is $\frac{48}{120} \times 100 = \frac{4800}{120} = \frac{480}{12} = 40\%$

(2) What fraction of pupils go on to Further Education?

Method:

The fraction of pupils that go on to further education is $\frac{32}{120} = \frac{8}{30} = \frac{4}{15}$**. The fraction representing this in its simplest form is** $\frac{4}{15}$

(3) What percentage of pupils is either employed or in apprenticeships? Give your answer to one decimal place?

Method:

Total number of pupils who are either in employment or apprenticeships = 25+15 =40, hence the percentage is $\frac{40}{120} \times 100 = \frac{4}{12} \times 100 = \frac{1}{3} \times 100 =$

33.3%

Example 3:

Key Stage 3 results in English in two adjacent areas A and B are shown in the pie charts below. Area A recorded the results of 840 pupils and area B recorded the results of 900 pupils in this subject. The pie charts show the percentages obtained in the appropriate levels at KS3. How many more pupils obtained level 7 in area B compared to those gaining level 7 in area A?

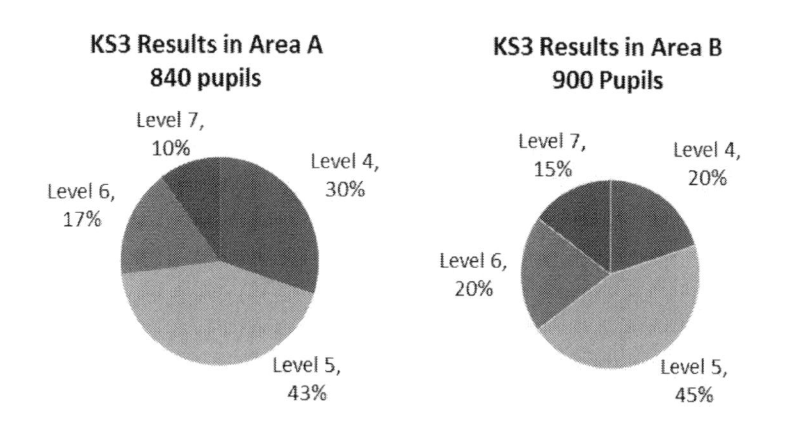

Method: The percentage of pupils who get level 7 in area A is 10%. This means 10% of 840 pupils get this level. 10% of 840 = $\frac{10}{100} \times 840 = 84$ pupils.

Similarly in area B, 15% of 900 pupils get level 7.

15% of 900 = $\frac{15}{100} \times 900 = 135$ pupils. This means area B has 135 – 84 more

pupils = 51 more pupils who get level 7 compared to A

Bar charts

Bar charts can be represented in columns or as horizontal bars. They can be either simple bar charts that show frequencies associated with data values or they can be multiple bar charts to allow for comparisons between data sets as shown below. The examples below illustrate some of the ways bar charts can be used to represent data.

Example 1: In a certain year 11 group the number of pupils who got a grade B or above in GCSE in Science, English, Maths and History was recorded as shown in the bar chart below.

(1) In which subject did the pupils get most B grades or above?

Answer: You can see from the column bar chart below that History was the subject where 40 pupils got a B grade or above, which is higher than any other subject

(2) What was the proportion of pupils who got B grade or above in History compared to the total? Give your answer as a fraction in its lowest terms

Answer: The number of pupils who got Grade B or above were 20 in Science, 35 in English, 25 in Maths and 40 in History. This means the total number of pupils who got Grade B or higher in these subjects = 120. Since 40 pupils achieved this level in History, the proportion in History compared to the total $\frac{40}{120}$ which simplifies to $\frac{1}{3}$

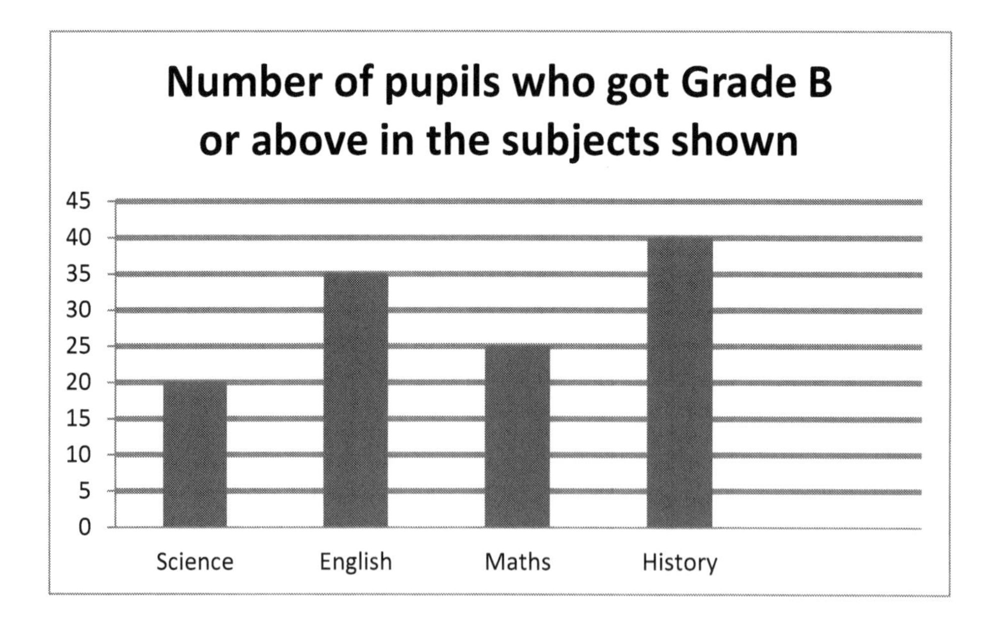

Example 2:

The bar chart below shows the amount of time in hours John, Bob and Bill spend surfing the web at weekends. What is the mean time per boy that is spent surfing the web at the weekend?

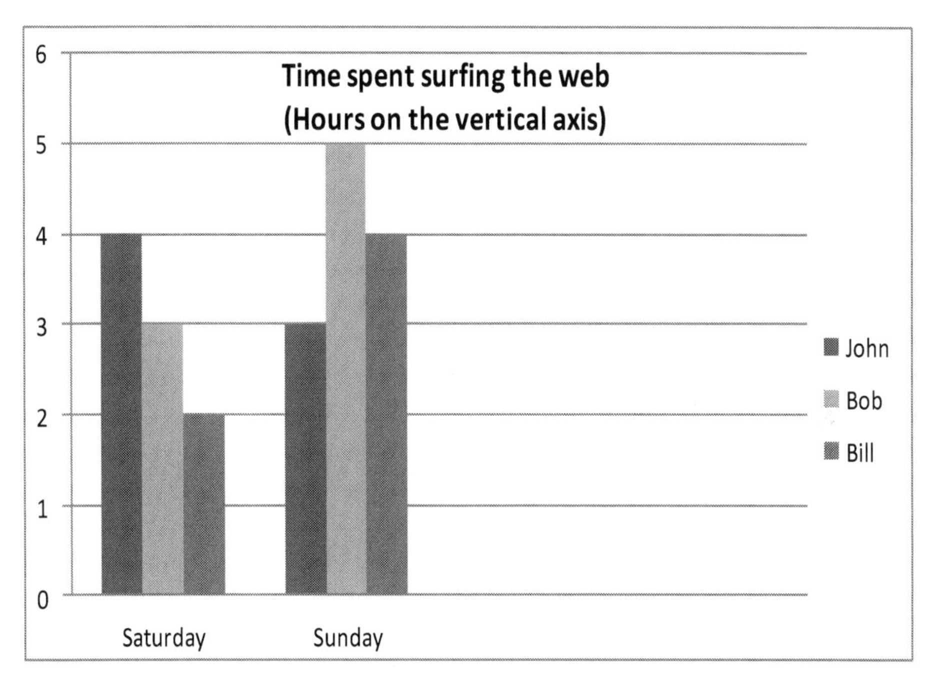

Method: John spends 4 hours on Saturday and 3 hours on a Sunday: a total of 7 hours

Bob spends 3 hours on Saturday and 5 hours on Sunday: a total of 8 hours

Similarly, Bill spends a total of 2 + 4 = 6 hours on a weekend

Total time spent surfing between the 3 boys on a week end is 7+ 8 + 6 =21hours

Hence the mean time spent per boy is 21 ÷ 3 =7 hours

Example 3:

Horizontal Bar Chart

In a school the percentage of girls who play different sports is recorded in a bar chart as shown below.

(1) If there are 560 girls altogether, how many girls play basketball?

(2) How many more girls swim than play hockey?

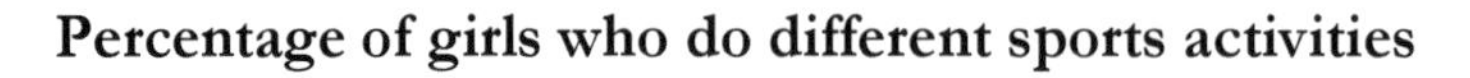

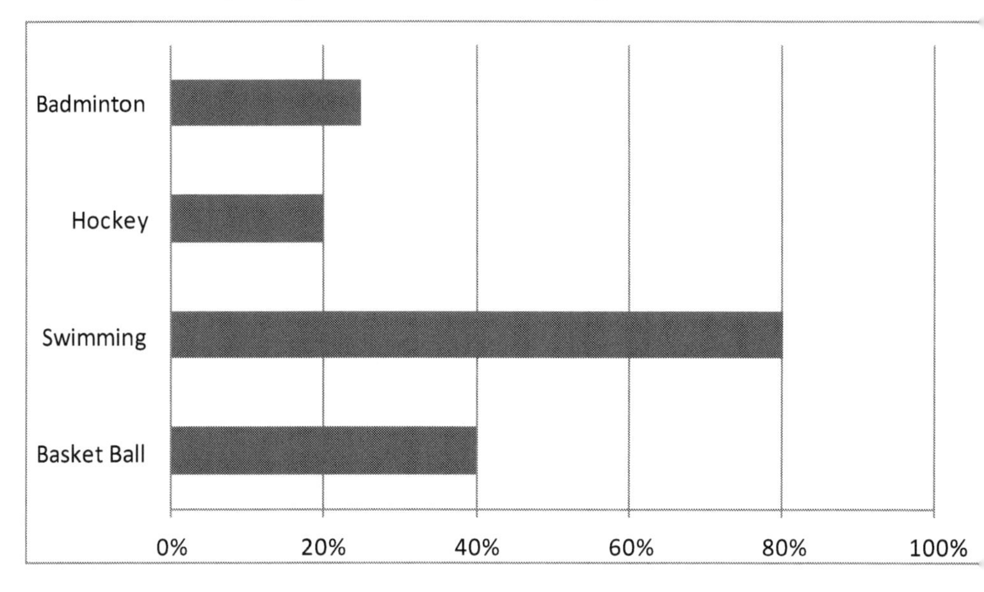

Method:

(1) From the bar chart you can see that 40% of the girls play Basket Ball. Since there are 540 girls altogether, this means 40% of 560 = 224 girls (10% of $560 = 56$, hence $40\% = 4\times56 = 4\times50 + 4\times6 = 200 + 24 = 224$)

(2) 80% of girls swim and 20% of the girls play Hockey. So 60% more girls swim compared to playing Hockey. 60% of 560 girls $=56\times6 = 336$. Hence, 336 more girls swim compared to playing Hockey

Example 4:

This composite bar chart below shows the percentage of pupils in a particular school who take and do not take additional lessons in music and maths respectively. What is the proportion of pupils who take extra music lessons? Give your answer as a fraction in its lowest terms.

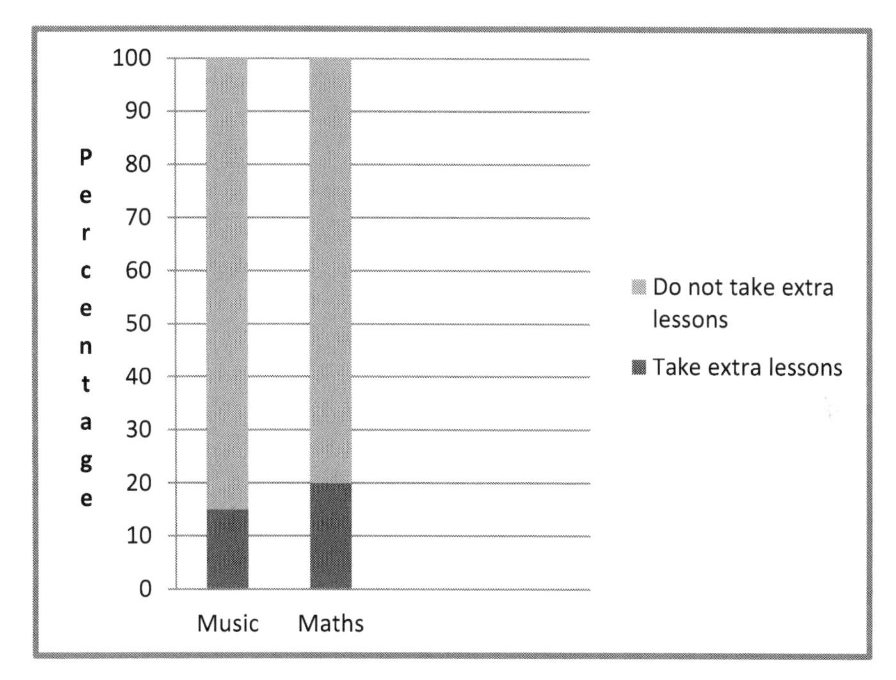

Method: The proportion of pupils who take extra music lessons is 15%. This is $\frac{15}{100}$ which simplifies to $\frac{3}{20}$. Hence $\frac{3}{20}$ of the pupils take extra lessons in music.

Summary of composite bar charts:

Although a composite bar chart consists of single bars, these bars are split into two or more sections. These sections show the frequencies of the appropriate categories. Frequency in the above example were the percentages of pupils and the categories in this case were those that took and do not take music and maths lessons respectively.

(Note the composite charts can be of different lengths for example if they are dealing with the number of pupils rather than percentages in the vertical column – they can also have more than two sections, for example music lessons taken by year 7, year 8, and year 9 would have 3 sections)

Histograms

Histograms are a kind of bar charts that represents frequencies. They are similar looking to bar charts but have some basic differences

(1) You need to remember that there are **no gaps between the bars**, as the data shown is continuous
(2) The bars can be either **equal widths or different widths** depending on the class interval
(3) The vertical axis can be used to estimate frequencies

Example

The following marks were obtained on a maths class test as shown by the table below. Show the information as a histogram

Marks	$10 \leq m < 20$	$20 \leq m < 30$	$30 \leq m < 50$	$50 \leq m < 70$
Frequency	4	6	14	2

Reading the marks in the table:

The class interval $10 \leq m < 20$ means the marks are between 10 and 20. more precisely, the table shows that 4 pupils got greater than or equal to 10 marks and less than 20 marks. $10 \leq m$ means m is greater than or equal to 10. Similarly $m < 20$ means m is less than 20.

Notice the class intervals (the widths) can vary. The histogram is shown below:

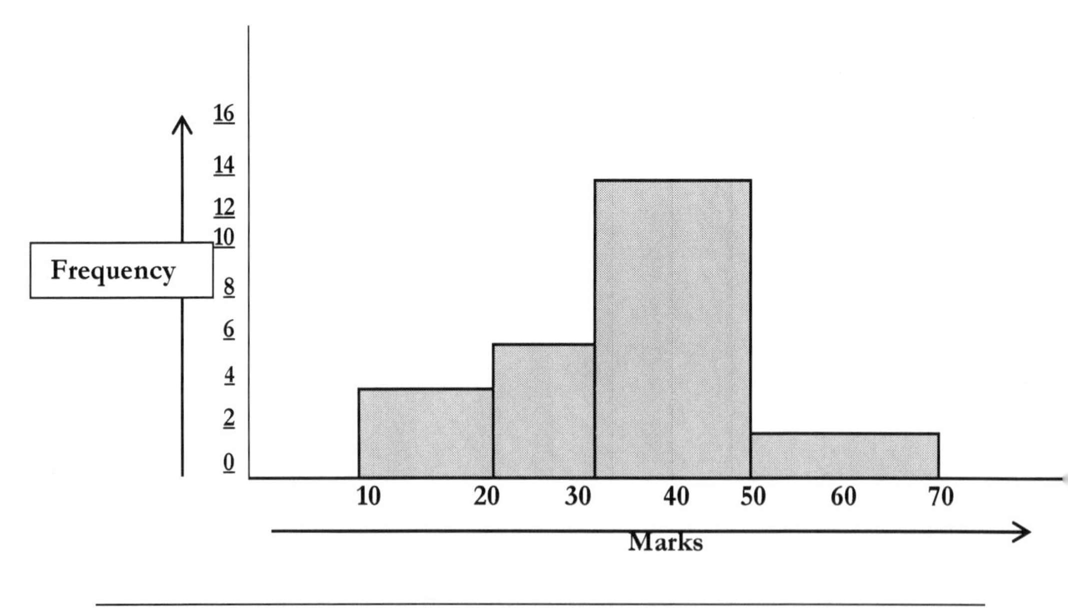

Scatter Graphs

Scatter graphs or diagrams are used to show the type of relationship between two variables, for example height and weight, maths and physics scores, reading scores and IQ and so on. It also gives information on the type of correlation between the two variables.

Positive Correlation

This means when the value of one variable increases so does the value of the other one as shown in the example below.

A certain number of pupils' test marks in maths and physics are plotted. (The straight line drawn is called the line of best fit). You can see in this case there is a positive correlation between Maths marks and Physics marks.

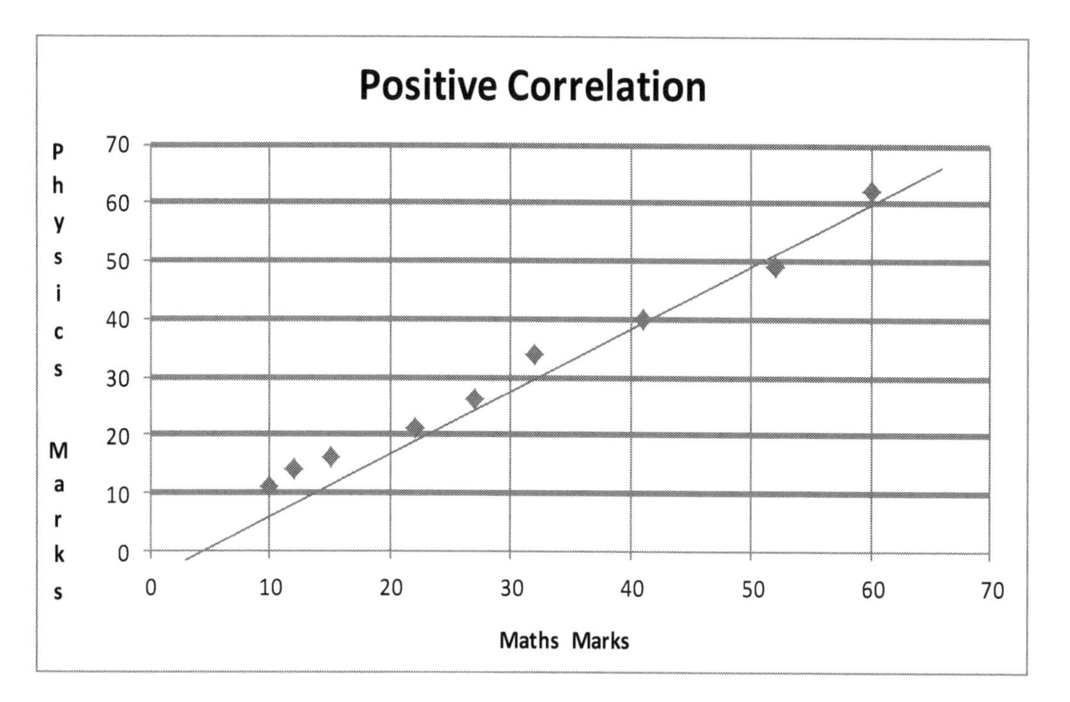

(1) How many pupils get more than 45 marks both in Maths and Physics?

Answer: 2 pupils get above 45 marks in both these subjects.

Method: look at the horizontal and vertical axis and draw an imaginary line at 45. Above 45 marks in both subjects you can see the record of two pupils.

(2) How many pupils get less than 20 marks in both Maths and Physics?

Answer: 3 pupils get less than 20 marks (using similar reasoning to the first answer)

Negative Correlation

This means that as one variable decreases in value the other one increases.

Example of Negative correlation:

This time some pupils' scores in Maths were plotted with their scores in History. It appears that in this particular case there was a negative correlation between Maths and History scores as shown below.

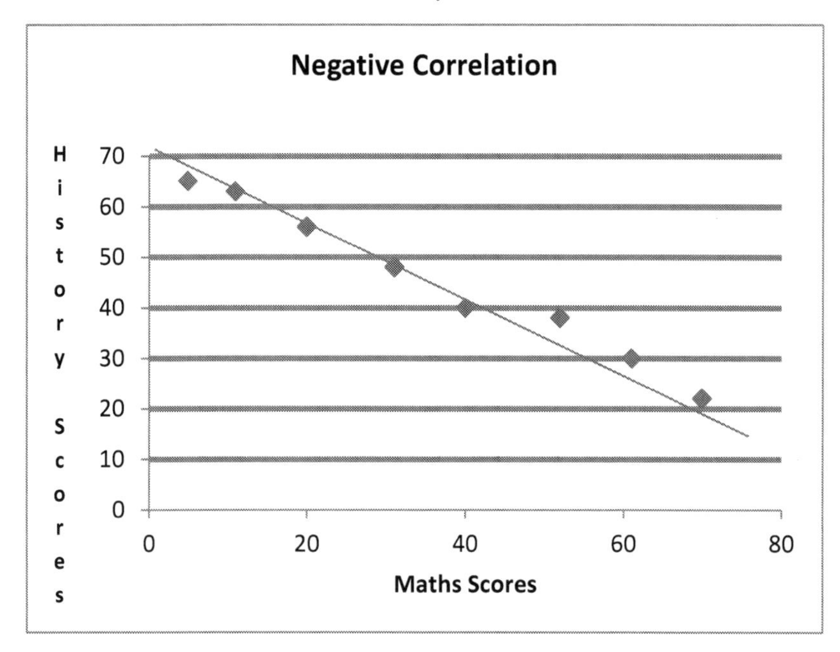

Another example of a negative correlation is the one between the price of a car and its age. As a car gets older its price is generally lower.

Zero Correlation

This is when there is no relationship between the two variables. The points are scattered all over the place so that we cannot really draw a line of best fit. For example consider the relationship between Science and English marks shown below in this particular example

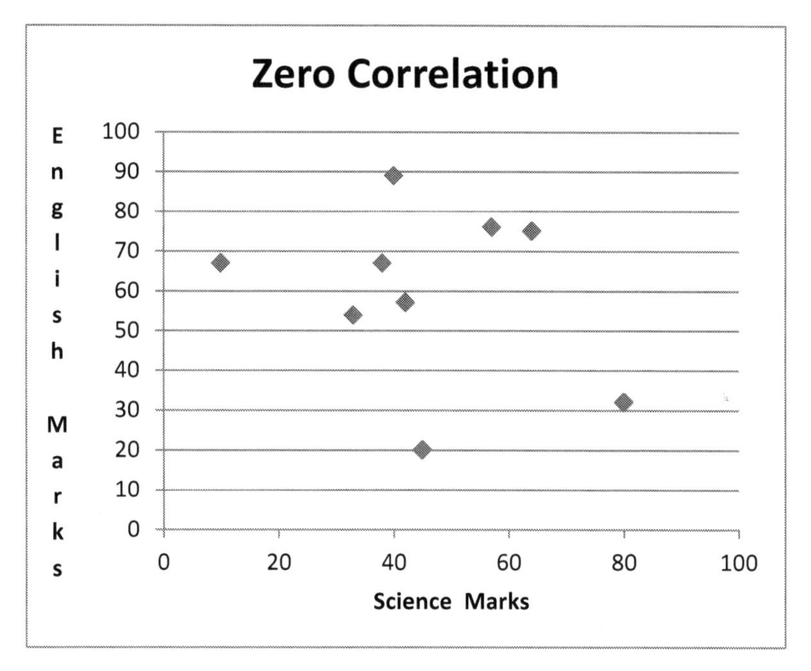

(1) In the example above what was the approximate mark in Science for the pupil who scored 20 marks in English?

Answer: 45 is the approximate mark in Science

Method: From the vertical axis, go along the horizontal line at 20 marks, this corresponds to approximately 45 marks in Science

It is worth remembering that we can have a weak positive correlation or a weak negative correlation. For example the percentage of men with grey hair is only weakly correlated with increasing age!

An example of scatter graph that shows a weak positive correlation, for example between Geography marks and History marks

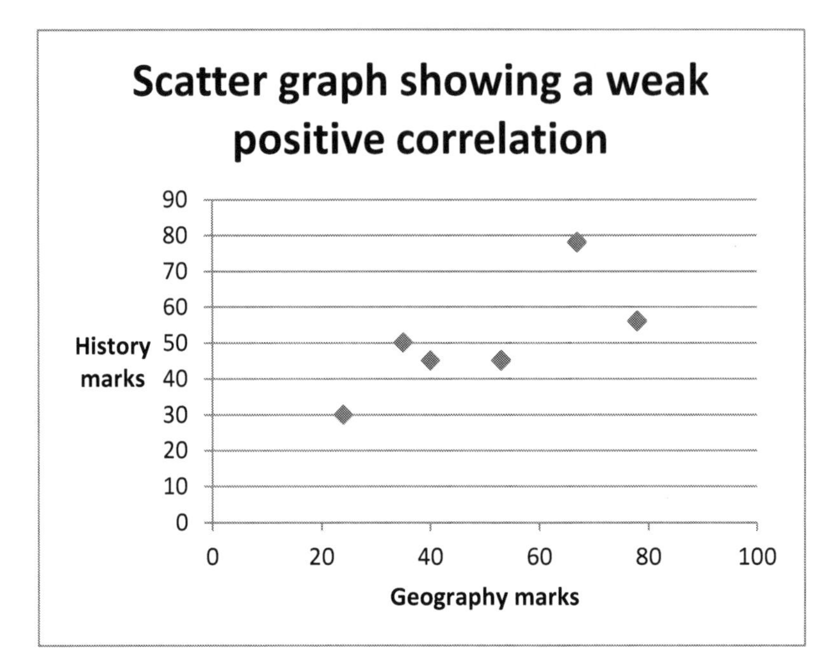

Finally, just for interest it is worth remembering the difference between correlation and causation. For example the incidence of heart attacks is correlated with high total cholesterol, but it is worth noting that many people with high cholesterol do not have a heart attack. In fact, the incidence of heart attacks is correlated with total cholesterol, LDL, triglycerides, obesity, body mass index and genetic factors. In other words, in this case the incidence of heart disease is multifactorial (has many factors that are responsible) and to draw a causal link from one factor may be erroneous. Another interesting and controversial issue is that more vaccinations have resulted in a higher incidence of autism! Unfortunately, as vaccination rates went up in the USA so did the rates of autism. This led some people to falsely attribute vaccinations as the cause of autism. Autism has gone up for other reasons, more awareness leading to more diagnostic cases, genetic factors which we were not previously aware of and so on. Often there are other 'variables' involved. Unfortunately, journalists and sometimes even scientists jump to 'causal' conclusions. In higher level statistics there are methodologies of significance testing which help ascertain whether a correlation is likely to be 'causal' or whether there could be other factors at play.

Line graph

A line graph is a way to represent two sets of related data. **It is often used to show trends**

Example1

The data below shows the percentage of pupils passing GCSE English at grades C or above from 2001 to 2006 in a particular school achieving. The same data can be shown as a line graph.

Year	2001	2002	2003	2004	2005	2006
% of pupils passing in English at grade C or above	26%	35%	45%	37%	48%	32%

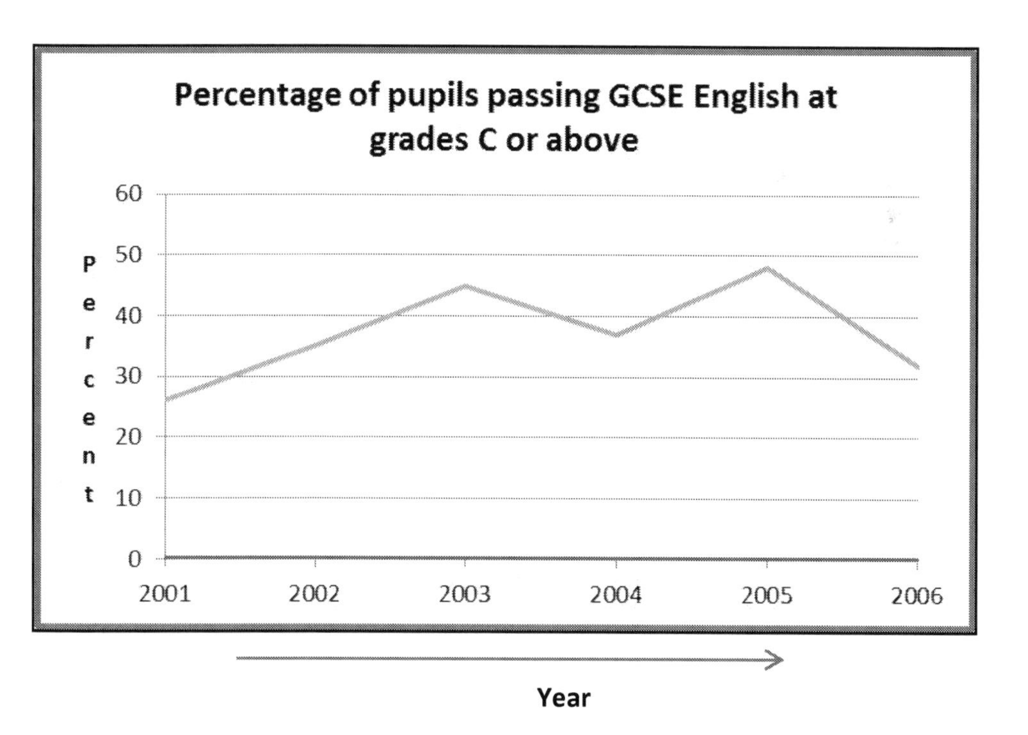

What was the change in success rate from 2004 to 2006?

You can see from the table as well as the graph that the success rate actually dropped from 37% to 32%. That is decreased by 5% points.

Example 2

In school A the percentage pass rates (Grades C or above) in English and Maths are both plotted in a line graph from 2001 to 2006. If 80 pupils took both English and Maths exams how many more pupils passed maths in 2006 compared to English in the same year?

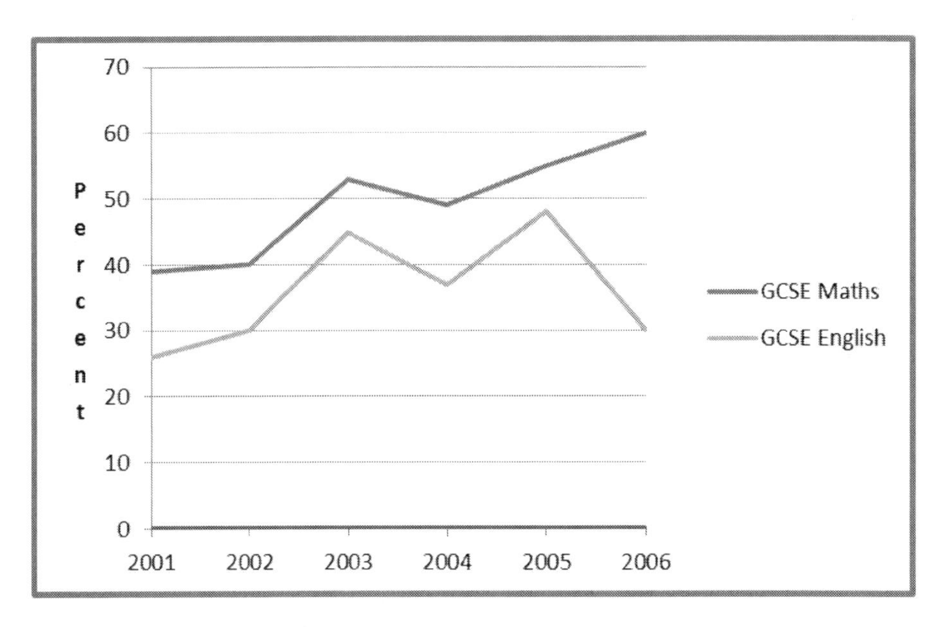

Method: From the line graph it can be seen that in 2006, 30% of the pupils passed their English GCSE at grade C or above. Similarly, in the same year 60% passed their maths GCSE at grade C or above. This means 30% more pupils passed Maths compared to English. Since 80 pupils took the exam this means 24 more pupils passed Maths compared to English.

(30% of 80 = 24)

Two way tables

These are used to compare data between two variables. For example comparing the different modes of transport (one variable) used by different schools, (second variable), say inner city schools and suburban schools. Example 1 demonstrates this.

Example 1:

		Method of Transport			
		Car	Bus	Walking	Other
Type of Schools	Inner City	28%	32%	24%	16%
	Suburban	62%	18%	12%	8%

From the data above you can see that 32% of children take the bus in Inner City schools compared to 18% who take the bus in suburban schools. Similarly, 62% of pupils in suburban schools arrive by car as compared to 28% in inner city schools. You can also compare other modes of transport between the two schools.

Example 2:

This second example shows a two way table comparing pupil results for GCSE History with GCSE Geography grades.

	History GCSE Grades								
Geography GCSE grades	A*	A	B	C	D	E	F	G	Total
A*		1	2	1					4
A									
B		1	2	3					6
C		2	3	4	2				11
D			2	3	1	1			7
E				2	1	1	1		5
F									
G									
Total		4	9	13	4	2	1		33

Typical questions

(1) How many pupils achieved a grade B in both History and Geography?

Method: Look down the vertical column (History) with Grade B and see where it crosses the horizontal row (Geography) with the same Grade. You can see that the cell that corresponds to both these being true is 2. This means that 2 pupils get a B in both History & Geography.

(2) How many pupils get a C in History?

Method: If you look down the vertical column (History) at grade C, you can see that the total pupils who get a grade C in History is 13.

(3) How many pupils achieved a grade C or above in Geography?

Method: Look at the horizontal row (Geography) and see how many got C, B, A and A* (that is C or above). If you look at the horizontal cells corresponding to these grades the totals are 11 for grade C, 6 for grade B, 0 for grade A and 4 for grade A*. This corresponds to a final total of 11+6+0+ 4 = 21. This means 21 pupils got a C grade or above in Geography.

(4) What is the percentage of pupils who got a grade B in History? Give the answer correct to 1 decimal place.

Method: Look down the vertical column (History) at grade B. The total number of pupils who got grade B in History is 9. Since 33 pupils took the exam altogether, the percentage of pupils who got a grade B in History is (9/33) × 100 = 27.27% which is 27.3% to 1 decimal place.

Box and Whisker plots

A box and whisker plot summarizes the **spread** of data. It shows the **median,** the **upper** and **lower quartile** as well as the **lowest** and **highest** value of the data set.

Remember the median is the middle value of the data set – this means half the data set is below and the other half above. The lower quartile is at the 25% of the data set. Similarly, the upper quartile is at 75% of the data set. The difference between the upper and lower quartile is called the Inter Quartile Range (IQR).

The lower quartile is often referred to as Q1 and the Upper quartile as Q3

The interquartile range is simply Q3 – Q1 that is Upper Quartile – Lower Quartile

However note that the range is simply the difference between the highest and lowest number. Consider some examples below:

Example 1:

The box and whisker diagram below represents the following marks in Science for 10 pupils. The marks are out of 40.

Pupil Marks In Science	14	22	12	16	34	6	28	30	8	5

(1) Let's find the median first. Arrange the marks in order so we get 5, 6, 8, 12, 14, 16, 22, 28, 30, and 34

The median is the middle number so it is between 14 and 16. (14+16)/2 = 30/2 =15

The upper quartile is three quarters of the way up (above this is the highest 25%). Another way of looking at this is that the upper quartile is the median value of the top half that is 16, 22, 28, 30 and 34. Which means the upper quartile is the middle number that is 28. Similarly the bottom half numbers are: 5, 6, 8, 12, 14 which means the Lower Quartile is the middle number of this data set which is 8.

The Inter Quartile Range (IQR) = Upper Quartile – Lower Quartile = 28 – 8 = 20

The range is however defined as the highest score minus the lowest score, so the range in this case is 34 - 5 =29

The box and whiskers diagram below illustrates this data as a visual representation.

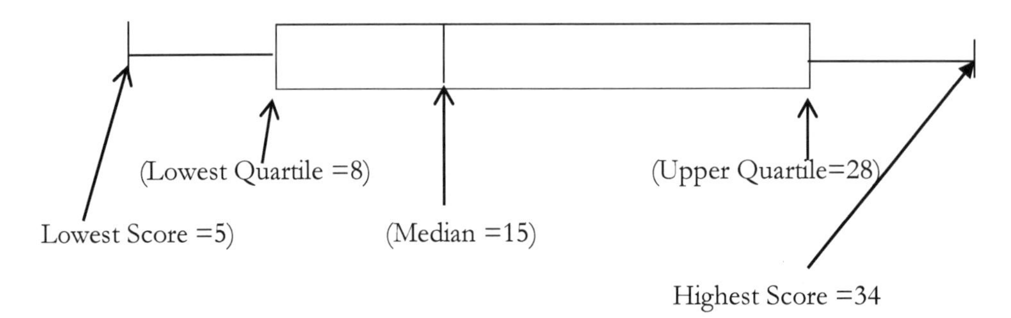

(Note the **ends of the Whiskers** denote the **lowest** and the **highest** scores. **The ends** of the **rectangle** represent the **lowest** and the **highest quartiles** as shown. Finally, the median is the **vertical line** inside the rectangle where it shows median =15)

Example 2:

Recent test marks in English, History and Geography for year 9 are compared by the three box and whiskers diagram shown below. The box and whiskers diagram shows the distribution of marks in English, History and Geography. So for example in English the lowest mark is 10, the lower quartile is 20, the median is 40, the upper quartile is 55 and the highest mark is 70. For History, the lowest mark is 15, the lower quartile is 30, the median is 50, the upper quartile is 60 and the highest mark is 75. Finally, for Geography, the lowest mark is 20, the lower quartile is 35, the median is 45, the upper quartile is 70 and the highest mark is 80.

Marks in English, History and Geography represented by respective box and whiskers diagrams

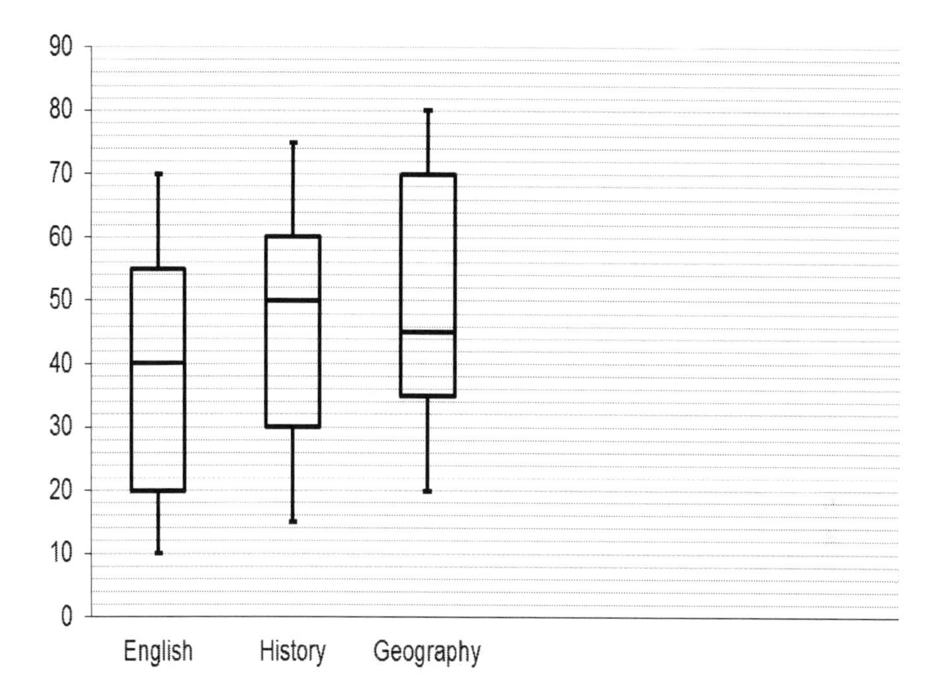

(1) Which subject had the biggest range?

Answer: All have the same range 60

In English the range is 70 -10 = 60, in History the range = 75 - 15 =60 and in Geography the range is 80 - 20 = 60

You can see that in this case all the subjects have the same range that is the difference between the highest and lowest mark in each subject is 60.

(2) In which subject did the top 25% perform best? **Answer:** Geography was the subject, where the top 25% got 70 marks or above

(3) What was the Inter Quartile Range in English? **Answer:** The inter quartile range is the difference between the upper and lower quartile, so in English the IQR (Inter Quartile Range) is 55 – 20 =35

(4) Which subject had the lowest mark? **Answer:** The lowest mark is in English which is 10 marks

(5) What subject had the lowest median mark? **Answer**: From the information given or by looking at the diagram you can see that the lowest median mark is in English

(6) Below what mark did a quarter of the English marks lie? **Answer:** Again from the information given or the box and whiskers diagram in English the appropriate mark was 20. That is 25% of the pupils in English got below this mark

Cumulative Frequency diagrams

Example 1:

A class test in maths was marked out of 70. The table below shows the distribution of marks among 20 pupils.

Marks (Max 70 marks)	Number of Pupils (frequency)	Cumulative Frequency (keep adding the frequencies)
1 - 10	0	0
11 - 20	2	0 + 2 = 2
21 - 30	3	2 + 3 = 5
31 - 40	5	5 + 5 = 10
41 - 50	5	10 + 5 =15
51 - 60	4	15 + 4 =19
61 - 70	1	19 +1 =20

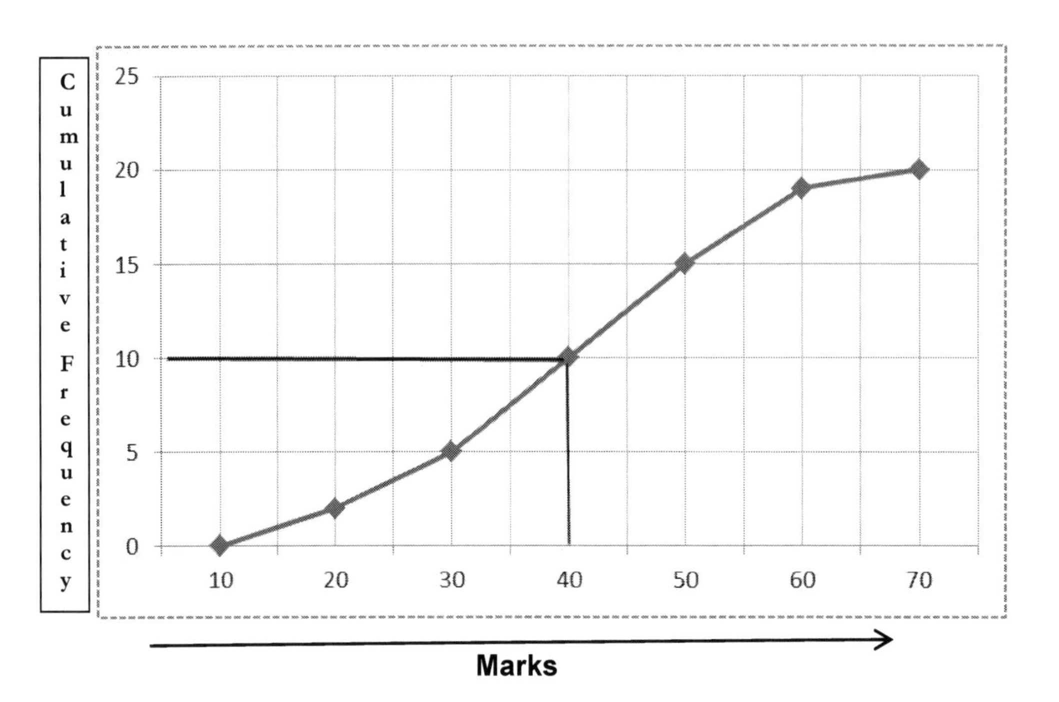

From the graph you can see the median (half way up the cumulative frequency is 10) this corresponds to 40 marks. This means half the students get up to 40 marks and half get over 40 marks.

(Just for your information when plotting the graph one plots the upper bound of the mark with the corresponding cumulative frequency value. So for example, in the mark range 21 - 30 you plot 30 (the upper bound) with 5)

Now, see if you can follow the answers to the questions below

(1) What mark corresponds to the lower quartile?

The lower quartile is a quarter of the way up the cumulative frequency axis. This corresponds to 5. The total cumulative frequency is 20, so a quarter on this axis is at 5. If you draw a horizontal line at 5 until it meets the curve and then draw the corresponding vertical line at this point it meets the horizontal line at 30. This means 25% of the pupils score 30 marks or below.

(2) What marks correspond to the upper quartile?

The upper quartile is three quarters of the way up the cumulative frequency axis. Three quarters of 20 (the total cumulative frequency) is 15. This corresponds to 50 marks on the horizontal axis.

This means that 25% of the pupils get more than 50 marks.

(3) What is the Inter Quartile Range (IQR)?

This is simply the difference between the Upper Quartile and the Lower Quartile.

IQR = Upper Quartile – Lower Quartile

IQR = 50 - 30 = 20

Summary

To find Median: From the vertical axis (which represents the cumulative frequency) go up to 50% or half way up this axis and draw a horizontal line to the cumulative frequency curve, then draw a vertical line at this point to meet the horizontal axis and read off the appropriate value

To find the Upper Quartile go up the 75% mark vertically (three quarters of the way up) and similarly read the corresponding value on the horizontal axis.

To find the Lower Quartile go up the 25% mark vertically (one quarter of the way up) and read the corresponding value on the horizontal axis

To find the Inter Quartile Range simply take the difference the Upper Quartile and the Lower Quartile.

Example 2:

The table below shows the distribution of hours of homework per week among Year 10 pupils.

Hours of Homework per week	Percentage of pupils	Cumulative Percentage
1	10	10
2	15	25
3	20	45
4	30	75
5	20	95
6	5	100

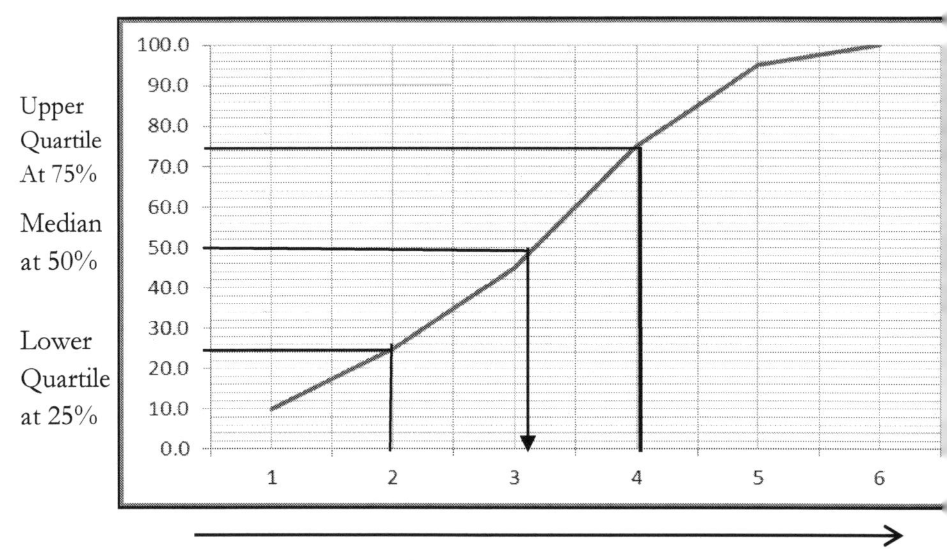

Hours of Homework

What is the median number of hours of homework done per week?

The median homework is approximately 3.2 hours per week (this means 50% of the pupils do 3.2 hours or less and 50% do more than 3.2 hours)

How many hours of homework approximately does the Upper Quartile represent?

Method: The upper quartile is at the 75% mark corresponds to 4 hours of homework per week

What are the minimum and maximum hours of homework done per week?

The minimum homework (the lowest value) is 1 hour per week, similarly, the maximum or the highest value is 6 hours per week

We can also represent this data as a box and whiskers diagram as shown below:

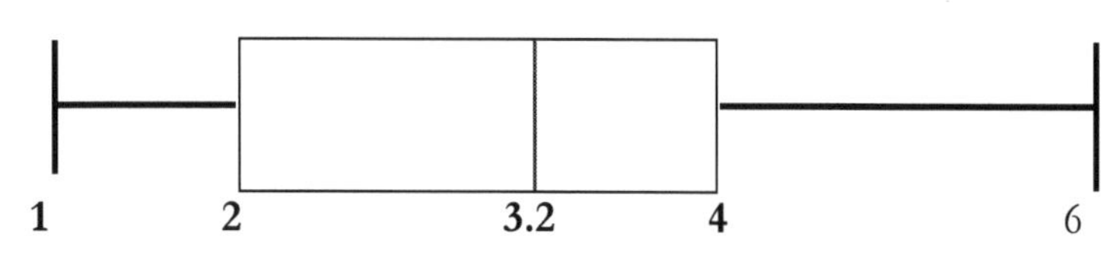

Median = 3.2 hours

Lower Quartile = 2 hours

Upper Quartile = 4 hours

Lowest value = 1 hour

Highest value = 6 hours

(The ends of the whiskers represent the highest and lowest values. The end of the bars represent the Upper and Lower Quartiles and the vertical line in the bar represents the median)

Finally, we can also work out the Inter Quartile Range and the Range if necessary

Interquartile Range = Upper Quartile – Lower Quartile = 4 – 2 = 2

Range = Highest Value – Lowest Value = 6 – 1 = 5

Chapter 7 Basic Algebra

The word 'algebra' comes from the Arabic al-jebr, which means 'the reuniting of broken parts'. By implication this means the equating of like to like.

In algebra we often use letters instead of numbers. There are some basic conventions and rules of algebra that you should be familiar with to progress in this subject. This chapter will be useful for you if you have forgotten your algebra.

If you see	We Mean
$x = y$	x equals y
$x > y$	x is greater than y
$x < y$	x is less than y
$x \geq y$	x is greater than or equal to y
$x \leq y$	x is less than or equal to y
$x + y$	the sum of x and y
$x - y$	subtract y from x
xy	x times y
x/y	x divided by y
$x \div y$	x divided by y
x^n	x to the power n
$x(x + y)$	x times the sum of x + y

Also note that:

$x(x + y) = x^2 + xy$

$x^2(x + x^2 + y) = x^3 + x^4 + x^2 y$

In general, $a \times a \times a \times a$ …….(n times) $= a^n$

You also need to know these algebraic rules for the multiplication and division of positive and negative numbers.

Multiplying positive and negative numbers.

(+) × (+) = + (a plus number times a plus number gives us a plus number)

(+) × (–) = –- (a plus number times a minus number gives us a minus number)

(–) × (+) = – (a minus number times a plus number gives us a minus number)

(–) × (–) = + (a minus number times a minus number gives us a plus number)

Dividing positive and negative numbers.

(+) ÷ (+)= + (a plus number divided by a plus number gives us a plus number)

(+) ÷ (–)= – (a plus number divided by a minus number gives us a minus number)

(–) ÷ (+) = – (a minus number divided by a plus number gives us a minus number)

(–) ÷ (–) = + (a minus number divided by a minus number gives us a plus number)

Summary: <u>For both multiplication and division, like signs gives us a plus sign and unlike signs gives a minus sign</u>

Also when adding and subtracting it is worth knowing that:

When you add two minus numbers you get a bigger minus number.

Example 1: – 4 – 6 = –10

When you add a plus number and a minus number you get the sign corresponding to the bigger number as shown below:

Example 2: +6 – 9 =–3, whereas, –6+9 =3

When you subtract a minus from a plus or minus number you need to note the results as shown below:

Example 3: 6 –(– 3) we get 6+3 =9 (since –(–3) =+3)

Example 4: 7 –(+3) we get 7 – 3 =4 (since –(+3) =–3)

In this case note that – (–) =+. Also, +(–) =– and –(+) =–.

Remember the BIDMAS rule you were introduced to earlier which specifies the rules concerning the order in which you carry out arithmetical operations:

Simplifying algebraic expressions

Example 1: Simplify $3x + 4x + 5x$

Method: We simple add up all the x's.

Hence we get $3x+4x+5x = 12x$

Example 2: Simplify $3x + 4x + 3y + 5y$

Method: Add up all the like terms.

So we get $3x+4x + 3y+5y = 7x + 8y$

(Notice we add up all the x's and then all the y's)

Example 3: Simplify $3m + 4y + 2m - 3y$

Method: as before, we add and subtract like terms.

Now $3m+2m = 5m$ and $4y-3y = 1y$ or just y.

So we can write $3m + 4y + 2m - 3y = 5m + y$.

Multiplying out brackets.

Example 1: Expand $3(2x + 5)$

Method: we multiply 3 by each term in the bracket. So we get $3 \times 2x + 3 \times 5$ which gives us $6x + 15$.

Example 2: Expand and simplify $3(2x + 5) + 4(2x+7)$

Method: Multiply 3 by each term in the first bracket then 4 by each term in the second bracket. The final step is to simplify by collecting up the like terms.

$3(2x+5) + 4(2x+7) = 6x+15+8x+28 = 14x + 43$

Notice the last step is simply adding $6x + 8x$ and then $15+28$.

Algebraic Substitution

This is the process of substituting numbers for letters and working out value of the corresponding expression.

Example 1: if $a = 5$ and $b=6$ work out $2a + 3b$

Method: Substitute numbers for letters and we get:

$2 \times 5+3 \times 6$

(Notice 2a means $2 \times a$ and 3b means $3 \times b$)

So, $2 \times 5 + 3 \times 6 = 10+18 = 28$

This means that $2a+3b = 28$

Example 2: If $m=7$ and $n=8$ work out $5m-3n$

Substituting numbers for letters we get:

$5 \times 7 - 3 \times 8 = 35 - 24 = 11$

So $5m - 3n = 11$

Simple Equations

Consider the following English statements and their mathematical equivalent:

English Statements	Algebra
Something plus five equals ten	$x + 5 = 10$
Something times two, plus five equals eleven	$2x + 5 = 11$
Something times three, minus five equals thirteen	$3x - 5 = 13$
Something divided by two equals three	$x/2 = 3$

Now consider solving these equations using a common sense approach.

Example 1: Something plus five equals ten. What is 'something'?

Clearly we need to add five to five to get ten. So 'something' in this case equals five.

Solving this by algebra can be very similar. As we saw, we can re-write the English statement above in algebra as follows:

$x + 5 = 10$ (notice, we are representing 'something' by x)

Now, if $x + 5 = 10$ clearly x (which represents 'something') is equal to 5.

So, $x=5$

Example 2: 'Something' times two plus five equals eleven. Find the 'something'.

We know that 'something' times two plus five equals eleven.

So the two times 'something' must equal 6. In which case 'something' must be 3.

Now consider the algebraic equivalent.

$2x + 5 = 11$

This means $2x = 6$

Which means $x = 3$

Now consider a more formal method.

Imagine an equation like a balance. Whatever you do to one side you must do to the other.

Example 3: Solve the equation $x + 5 = 10$

Subtract 5 from both sides

So, $x = 5$

However, we can also use the method of taking inverses.

The rules are: When something is added to the x-term subtract, when something is subtracted from the x-term then add. When x, is multiplied, by a number we divide. Finally, when the x-term, is divided, by a number we multiply.

Example 1:

Two pupils collect the same amount of money each for a charity. The teacher adds £5.50 and the total amount they collect is £11.70. How much does each pupil collect?

To solve this algebraically, let the amount each pupil collect be x. This means 2x +£5.50 =£11.70 (2 times the amount each pupil collects plus the £5.50 the teacher contributes = £11.70)

In the equation, $2x + 5.50 = 11.70$, we now have to Subtract 5.50 from both sides

So we are left with $2x = 6.20$

Now divide both sides by 2(that is, take the inverse of ×2)

So, x =£3.10, This means each pupil collects £3.10

Example 2:

John has £22 more than Brian. Altogether they have £68. How much do they each have?

Let the amount Brian has be £x

Since Brian's amount plus John's amount = £68, we can write algebraically that $x + x + 22 = 68$.

Simplifying this expression we get $2x + 22 = 68$

Subtracting 22 from both sides we get:

$2x = 46$, now divide both sides by 2. We get $x = 23$

Hence Brian has £23, and John has £45

Example 3: (Simultaneous equations where you have two unknowns in this case)

The sum of two numbers is 30. Their difference is 12. What are the two numbers?

Method: Let one of the numbers be x and the other y.

So we can deduce that (1) $x + y = 30$ and (2) $x - y = 12$.

If you add the left **and** right hand sides of the two equations then we can eliminate y. That is $x + y + x - y = 30 + 12$. This simplifies to $2x = 42$, or $x = 21$. Now that we have found x, we can substitute for x in the first equation. i.e. (1) to give us $21 + y = 30$. By subtracting 21 from both sides we can find that $y = 9$. Hence the two numbers are 21 and 9.

On- screen questions – Mock Test 1

Question 1

Test marks in Maths, Science and English for year 11 are compared by the three box and whiskers diagram shown below. Marks are shown on the vertical axis. Indicate all true statements:

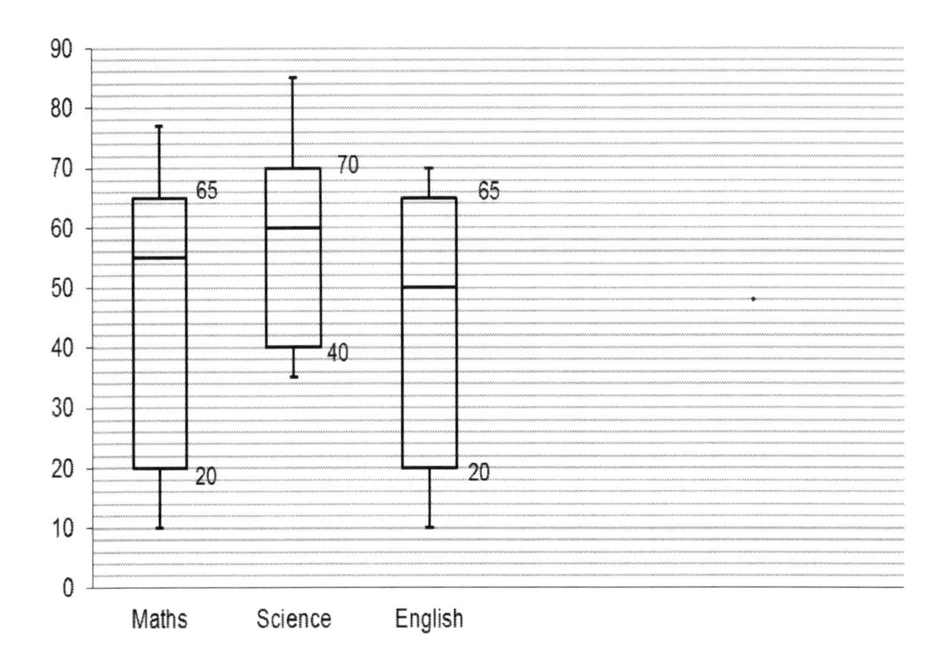

(1) The interquartile range for Maths is higher than that for English ☐

(2) The difference between the median marks for Maths and Science was approximately 5 marks ☐

(3) The highest mark was in Science ☐

Question 2

The head of English orders some books and DVD's for a year 10 group. The orders are for 92 set books at £1.65 each, 220 exercise books which come in packs of 10 at £4.60 per pack, and 8 Shakespeare DVD's at £5.75 each. There is a school discount of 12.5% on the total order. Calculate the total amount the order cost after the discount. Give your answer correct to 2 decimal places.

Total cost is £ ☐

Question 3

The set of data below shows the result in a year 10 Geography test for 96 pupils. The marks are out of 10. The teacher wants to find the mean mark for this test. Give your answer to 1 decimal place.

Marks in Geography Test	No of pupils	No. of pupils × Geography marks	
1	1	$1 \times 1 = 1$	
2	5	$2 \times 5 = 10$	
3	12		
4	22		
5	27		
6	18		
7	7		
8	3		
9	1		
10	0		
Totals	96		

The mean mark is:

Question 4

A Mock exam in English consists of two papers. The paper is out of 60 and has weighting of 65% given to it. The second paper is out of 80 and has a weighting of 35% given to it. A pupil gets 42 in the first paper and 48 in the second paper. What is the pupil's final percentage score?

Weighted score is:

Question 5

The two way table shown compares pupils' results for GCSE Maths with GCSE English grades.

	Maths GCSE Grades								
English GCSE grades	A*	A	B	C	D	E	F	G	Total
A*			3	1					4
A		1	1	3					5
B		3	2	5					10
C		1	5	8	2				16
D			3	3	1				7
E				2	1				3
F									
G									
Total		5	14	22	4				45

Indicate all true statements

(1) The number of pupils who achieved a grade A in both Maths and English is 3

(2) The number of pupils who got a grade C in English is 16 ☐

(3) The percentage of pupils who got a B in Maths is approximately 66.7% ☐

Question 6

A walking trip was organized. The map showed a scale of 1:100000. The teacher planned out the route as follows:

Start from A and going to B total distance on the map =2.7cm

From B to C the distance on the map was 3.2 cm

Finally the distance from C to D was 8.2 cm

What was the distance in Kilometres from A to D?

Answer ☐ Km

Question 7

The pie chart below shows the number of pupils who got a Grade C or better in English in four different schools.

Indicate all true statements below:

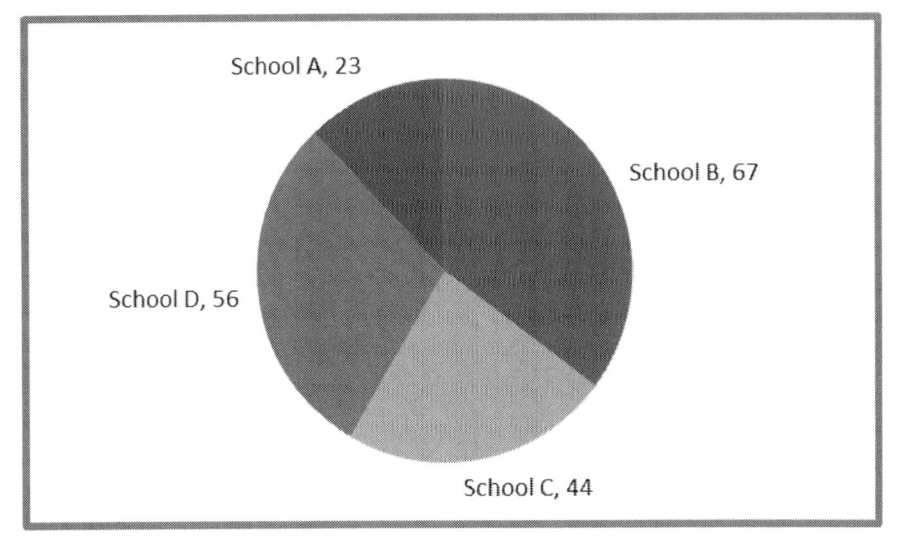

(1) The percentage of pupils who got Grade C or better in English in School C compared to all the schools combined was:

Approximately 23.2% ☐

(2) The proportion of pupils getting a grade C or above in English at School D compared to all schools was $\frac{28}{95}$ ☐

(3) The total percentage success at Grade C or above in English at School A and School B combined compared to all schools was approximately 47.4% ☐

Question 8

A teacher plans a day trip to Brighton for her year 9 pupils. She calculates that the distance to Brighton from her school is 180 kilometres. She estimates that the average speed of the coach will be 45 miles per hour. She plans to be in Brighton at 16:00.

What time should she leave the school to arrive there at her planned time? Give your answer using the 24 –hour clock

Answer 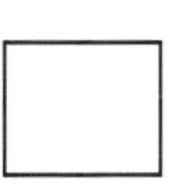

Question 9

Pupils who succeeded in getting English GCSE at Grade C or above were analyzed from 2005 to 2011

Indicate all true statements

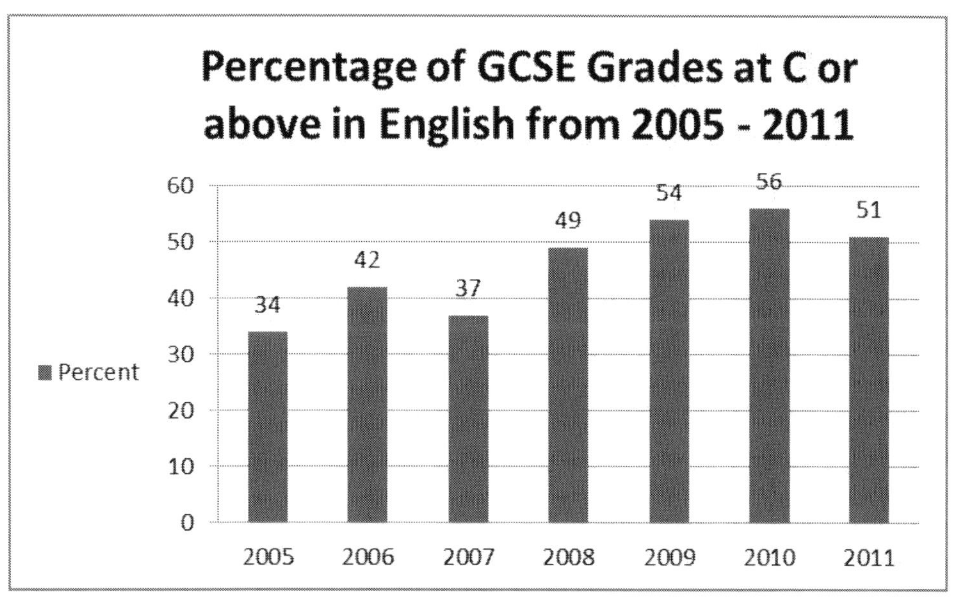

(1) The percentage of GCSE at grades C or above increased every year

(2) The mean percentage from 2008 to 2011 was 52.5%

(3) The increase in percentage points from 2005 to 2011 was 27%

Question 10

A teacher wanted to compare the progress of 8 pupils across two tests and see who had increased by at least 10 percentage points. The first test was out of 40 and the second test was out of 50.

Click the letters in the table for the pupils that meet this target.

Pupils	Test1 (marks out of 40)	Test 2 (marks out of 50)
A	22	32
B	25	32
C	17	27
D	25	35
E	19	26
F	12	20
G	25	34
H	30	47

Question 11

The head of maths created the following table showing the number of pupils in each year group who had additional maths tuition.

What is the percentage of pupils in all the year groups combined that are having additional tuition. Give your answer rounded to a whole number.

Year Group	Number of pupils	Number of pupils receiving additional maths tuition
7	98	12
8	103	16
9	110	15
10	97	12
11	101	11

Answer: 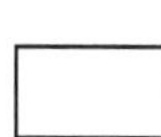

Question 12

English GCSE grades were recorded for 40 pupils. The data is shown on a cumulative frequency diagram below.

Indicate all true statements below:

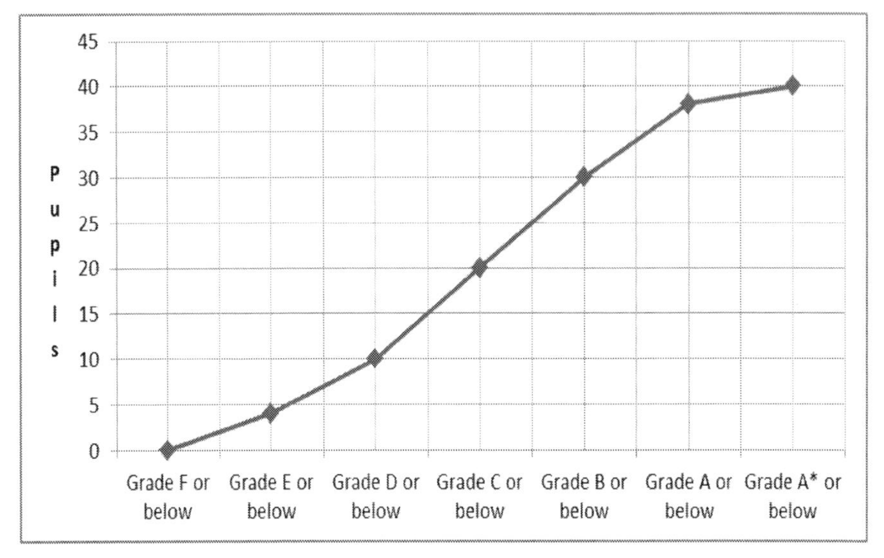

(1) The Median grade for the 40 pupil's GCSE grade is grade C or below ☐

(2) 30 pupils get above grade D ☐

(3) 30 pupils get above grade B ☐

Question 13

A teacher wants to convert a temperature from Celsius into Fahrenheit

The formula for converting the temperature from Celsius to Fahrenheit is given by:

$F= \frac{9}{5}C +32$ (where C is the temperature in degrees Celsius). If the temperature is 22 degrees Celsius what is the equivalent temperature in Fahrenheit?

Answer: 

Question 14

A teacher represents the relationship between marks in a maths test and a science test by the scatter graph shown below. The Maths marks are out of 60 and the Science marks are out of 10. Indicate all true statements below:

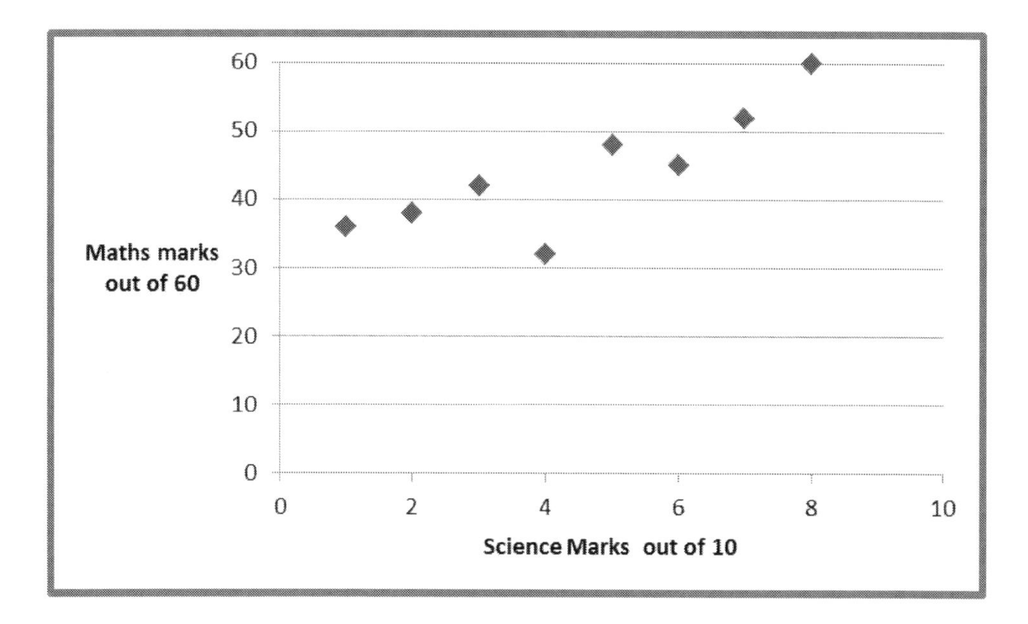

(1) The correlation between Maths and Science marks in this test was negative. ☐

(2) In general if a pupil's mark was high in Science it was also high in Maths ☐

(3) When the pupil's mark in Science was 8, the corresponding mark in maths was also 8 ☐

Question 15

A teacher in a Junior- school records the reading age for a group of 7 pupils over a six-month period. In each month the recording done is that of the reading age minus the actual age for each pupil.

Point and click on the pupil(s) that show a consistent trend of improvement every month in the difference between reading age and their actual age over a six-month period. (In this case instead of pointing and clicking, simply circle the appropriate pupil(s) in the first column)

	Reading Age minus actual age (months)					
Pupil	Month 1	Month 2	Month 3	Month 4	Month 5	Month 6
A	3	4	5	5	6	7
B	2	3	5	6	6	7
C	-5	-4	-3	-2	-1	0
D	1	2	2	2	3	3
E	-7	-5	-3	-1	1	3
F	4	5	6	6	6	7
G	-8	-7	-5	-5	-4	-3

Question 16

Two teachers take 12 pupils to France on a cultural visit. The exchange rate at the time they go is £1 = 1.3 euros. They change £50 for each pupil and £100 each for themselves before going to France. When they come back the exchange rate is £1 = 1.25 euros. If they convert all their Euros left back to pounds they get back £28. How much in Euros did they spend altogether whilst in France?

Answer

Answers – On screen Test 1

Question 1

(1) False – since the interquartile range for Maths is UQ- LQ = 65-20 =45 and likewise for English it is 45

(2)True - the median marks for maths = 55 and the median for science =60, so the difference =5

(3) True- The highest mark was 85, which was in Science

Tip: Remember that the Interquartile range is the difference between the Upper & Lower Quartiles. Also the bottom end of the box or bar represents the Lower Quartile and the Upper end represents the Upper Quartile. The dark horizontal line in the box represents the median and finally the lowest and highest marks are represented at the end of the whiskers)

Question 2

Answer = £261. 63

Method: 92 set books @ £1.65 each = 92X 1.65 = £151.80, 22 packs of exercise books @ £4.60 per pack = 22 X 4.60 = £101.20 and 8 DVD's @ 5.75 each = 8 × 5.75 = £46. Total cost before discount = £299. Since the school discount is 12.5%, the actual price is £299 – 12.5% of £299 or 87.5% of £299 = £261.63

Tip: If you discount something by 10%, you are effectively paying 90%. So if you discount something by 12.5%, you are effectively paying 87.5%. Simply work out 87.5% of £299 on a calculator.

Question 3

Mean = 460/96 = 4.8 (to 1 d.p.)

(Reminder: The mean = sum of all (the frequency × marks) ÷ by the total frequency

Question 4

Answer = 66.5%

Method: In the first paper the pupil scores $\frac{42}{60} \times 100 = 70\%$, in the second paper the pupil scores $\frac{48}{80} \times 100 = 60\%$. Now apply the appropriate weightings. So the total for both papers is: $0.65\times70 + 0.35\times60 = 45.5 + 21.0 = 66.5\%$, (Note that 65% of 70 = $\frac{65}{100} \times 70 = 0.65 \times 70$)

Question 5

(1) False – The number of pupils who obtained both grade A in Maths and English is 1

(2) True- Look across the horizontal column down English at Grade C, the total = 16

(3) False – the percentage of pupils who got a grade B in Maths = $(14/45) \times 100$ = approx. 31.1%

(Reminder: In a two-table go down vertically to obtain the appropriate variable (in this case GCSE maths marks – and similarly go across the horizontal table for the English GCSE grades)

Question 6

Answer = 14.1 km

Method: The total distance on the map = (2.7 + 3.2 + 8.2) cm = 14.1cm on the map

14.1cm using a scale of 1:100000 = 1410000 (multiply 14.1 by 100000)

Now 1410000 cm = 14100 metres (divide 1410000 by 100 to get answer into metres first)

Finally, $14100 \div 1000$ (to convert to Km) = 14.1 km

Question 7

Answers

(1) True – total number of pupils represented by the pie chart is 190 the number corresponding to School C is 44. Hence the percentage of pupils who got Grade C or better in English in School C = $\frac{44}{190}\times100 = 23.2\%$

(2) True – the number of pupils corresponding to School D is 56, hence the proportion is $\frac{56}{190}$, (dividing numerator and denominator

by 2, this simplifies to $\frac{28}{95}$

(3) True – the total number of pupils corresponding to Grade C or above in English at School A and School B = 90, so the corresponding percentage = $\frac{90}{190}\times100 = 47.4\%$ approximately

Question 8

Answer: 13:30

Method: The distance is 180 km which in miles is $180 \times \frac{5}{8} = 112.5$

miles. Since the average speed is in 45 miles per hour the time taken is 112.5/45 =2.5 hours or 2 hours and 30 minutes.

Since the teacher plans to be in Brighton at 16:00 simply subtract 2 hours and 30 minutes from this time to give 13:30

Question 9

(1) False – you can see from the bar chart that there were some years where the performance fell. For example from 2006 to 2007

(2) True – the mean percentage from 2008 - 2011 = $\frac{210}{4}$ =52.5%

(Note 210 is the total between 2008 – 2011)

(3) False – The increase in percentage from 2005 to 2011 was from 34 to 51 = 17%

Question 10

Answers:

The pupils who increased their marks by 10% points or more were C, F and H

Method: First work out each mark as a percentage as shown then see who had a 10% increase from Test 1 to Test 2

Pupils	Test1 (marks out of 40)	Test 2 (marks out of 50)
A	22 = 55%	32 = 64%
B	25 = 62.5%	32= 64%
C	17 = 42.5%	27 =54%
D	25 =62.5%	35 = 70%
E	19 =47.5%	26 =52%
F	12 =30%	20 =40%
G	25 =62.5%	34 =68%
H	30 =75%	47 =94%

Question 11

Answer = 13%

Method: Total number of pupils in all the year groups = 509, total number of pupils receiving additional maths tuition = 66. Hence percentage having additional tuition

$=\frac{66}{509} \times 100 = 12.9666\% = 13\%$ (rounded to a whole number)

Question 12:

(1) True – for the median go half way up the vertical axis (i.e. to 20, since the total number of pupils is 40), draw a horizontal line at 20, and where it crosses the cumulative frequency curve draw a vertical line to the horizontal axis. This gives you the appropriate Grade.

(2) True – Since 10 pupils get Grade D or below, this means 30 pupils get above grade D

(3) False – Since the total pupils who get Grade B or below is 30, the numbers that get above this Grade is only 10.

Question 13

Answer: 71.6 ° F

Method: In the formula $F=\frac{9}{5}C+32$, substitute 22 in place of C. So we get

$$F=\frac{9}{5}\times 22+32=39.6+32=71.6^{\circ}\ F$$

Question 14

(1) False – when both values tend to go up, we call this a positive correlation not negative correlation.

(2) True – since the relationship is positively correlated

(3) False –when the pupil's science score was 8, the maths mark was approximately 55

Question 15

Answer: Pupil C & E- since for C the reading age consistently goes up by one month, every month over this time period and for E it goes up two months every month in this time period.

Question 16

Answer: 1005 Euros were spent

Method: 12 pupils @ £50 each and 2 teachers @ £100 = total of £800, If £800 is converted to Euros at £1 =1.3 Euros, this means £800 = 1040 Euros. Finally, if they get back £28 and the exchange rate at the time was £1 = 1.25 Euros, then £28 = 35 Euros (since 28÷1.25), this means the amount of Euros spent = 1040 – 35 = 1005

On screen questions – Mock Test 2

Question 1

A teacher is planning a trip to the theatre. The price of each ticket is £4.85. There are 22 pupils and 4 teachers who go on this trip. The cost of return travel by coach for all the people is £48. What is the total cost of this outing?

Answer: 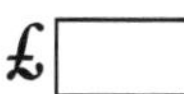

Question 2

The graph below shows the percentage of pupils achieving level 5 at Key Stage 3 Maths from 2006 to 2011.

Indicate all true statements

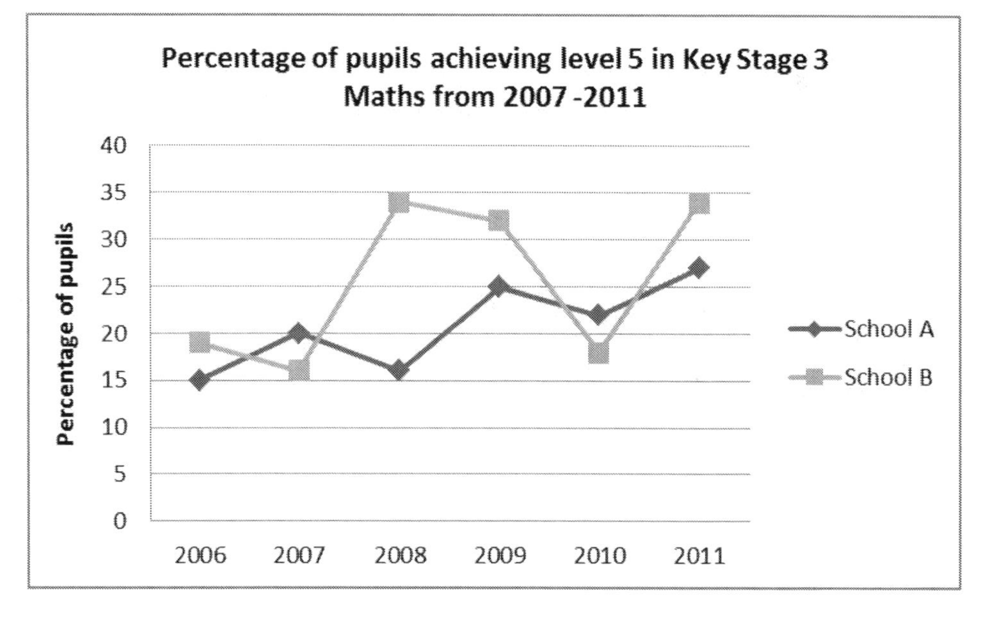

(1) School B outperformed School A in every year except 2010 ☐

(2) School A showed better results than School B in 2007 and 2010 ☐

(3) There was no difference in results between the two schools in 2006 ☐

Question 3

A school parents evening starts at 16:45. A teacher has 18 appointments of 14 minutes each. There is also a 30 minute break. When did the parents evening finish? Give your answer in the 24 hour clock.

Answer: [ ]

Question 4

English marks for year 11 over two years were summarized by a box and whiskers diagram as shown below: Indicate all true statements

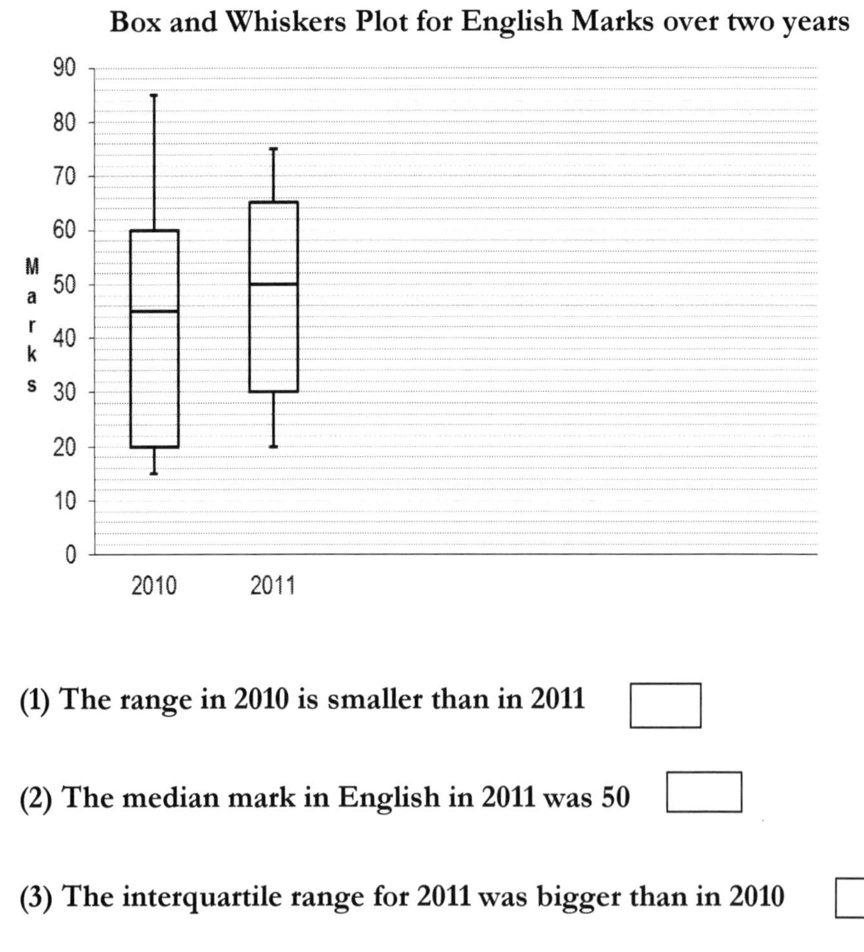

(1) The range in 2010 is smaller than in 2011 [ ]

(2) The median mark in English in 2011 was 50 [ ]

(3) The interquartile range for 2011 was bigger than in 2010 [ ]

Question 5

A pupil goes to Switzerland on a school trip and converts £80 into Swiss Francs (CHF)

at an exchange rate of £1 = 1.5 CHF. He spends 88 CHF whilst in Switzerland. When he returns to the UK he exchanges the remaining amount at an exchange rate of £1 = 1.55 CHF. How much does he get back in English money? Give your answer to the nearest penny

Answer

Question 6

On a school trip to Paris, the teacher plans an excursion from Paris to Versailles and back. She wants to be in Versailles by 15:00. The distance between their Paris accommodation and Versailles is 30Km. Because of anticipated traffic the coach they hire can only travel at an average 25 miles per hour. Using the approximation of 5 miles = 8 kilometres, what is the latest time she should plan to leave Paris?

Answer:

Question 7

In a certain area in London, a group of schools had a total of 850 pupils who obtained at least a level 5 at Key Stage 3 in Maths, Science, Geography and English. The percentage of pupils in this group who achieve at least a level 5 by gender in the appropriate subjects is given in the bar chart below.

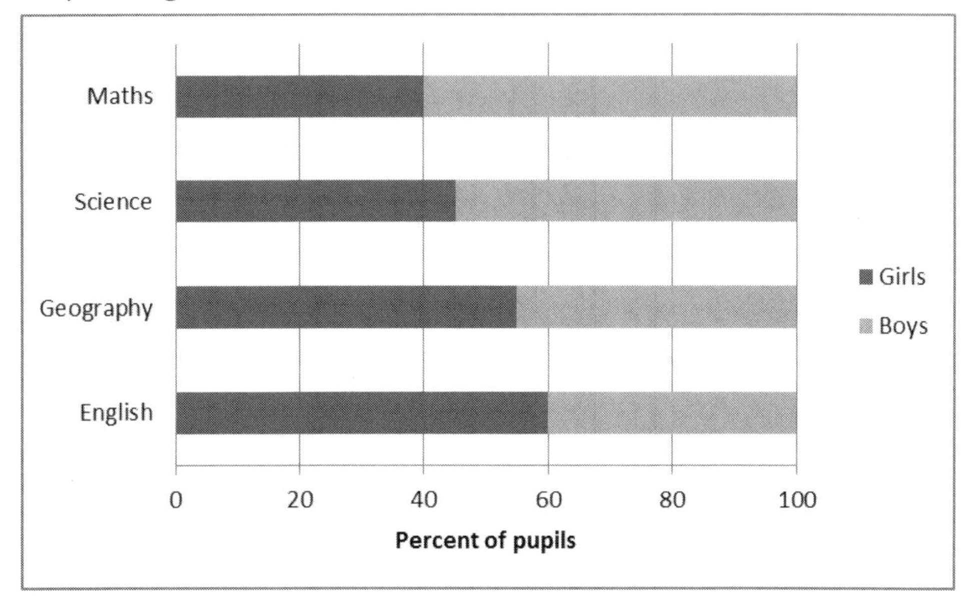

What was the total number of boys in this group of schools who got at least a level 5 in English?

Answer [ ]

Question 8

A teacher tells pupils that there will be a test in 4 weeks' time. She asks them to record their revision time over this period. Finally she shows the pupils the results of the maths test in relation to the time they spent revising in a scatter graph.

Indicate all true statements

Scatter graph showing revision time in hours and resulting Maths marks

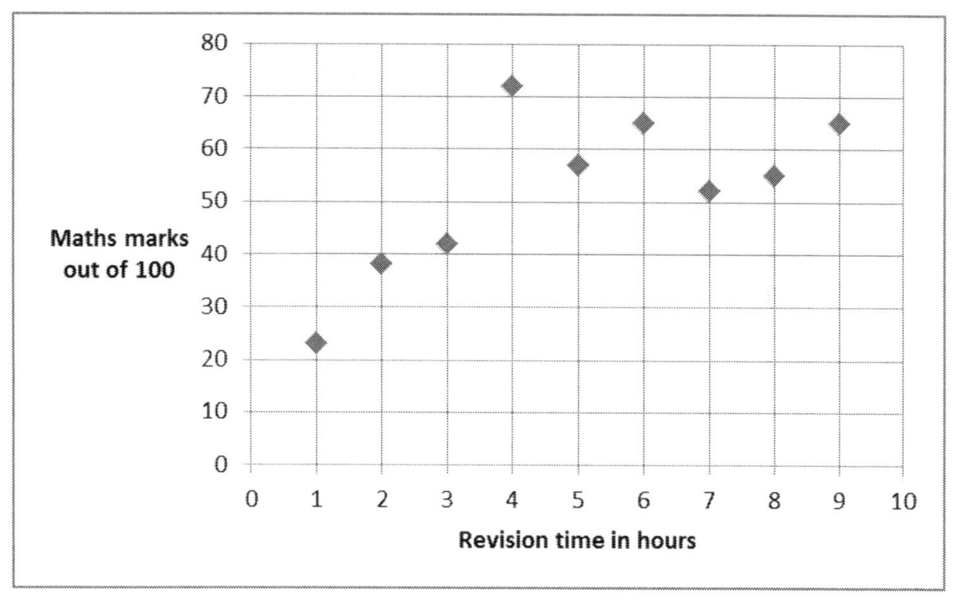

(1) The range in revision time is 8 hours ☐

(2) The number of pupils who get 50 marks or more in the test is 7 ☐

(3) All pupils who revise for 3 hours or less fail to get 50% or more ☐

Question 9

The percentage of pupils gaining level 5 or above at Key Stage 3 in six neighbouring schools is shown in the table below.

Point and click on the schools that show a continual trend of improvement every year. (In this case mark an 'X' on the right hand column for the appropriate schools)

School	2006	2007	2008	2009	2010	2011	Mark 'X' for School(s) that show continual improvement
A	23.5	23.7	23.2	24.2	24.1	24.6	
B	26.6	26.9	27.2	27.5	27.8	28.1	
C	16.7	18.0	19.0	20.0	19.5	21.1	
D	16.6	17.2	17.1	18.6	19.2	19.8	
E	28.3	29.3	29.1	31.2	33.2	32.1	
F	34.1	33.9	40.1	40.3	40.1	40.6	

Question 10

In a year 11 group, there are 54 pupils altogether of which 24 are girls. Music lessons are taken by $\frac{1}{5}$ of the boys and $\frac{1}{4}$ of the girls.

What is the total number of pupils from the year 11 group that take music lessons?

Answer [ ]

Question 11

In a certain local authority the percentage of teachers and the years of teaching service completed before retirement was plotted in a cumulative frequency diagram.

Indicate all true statements.

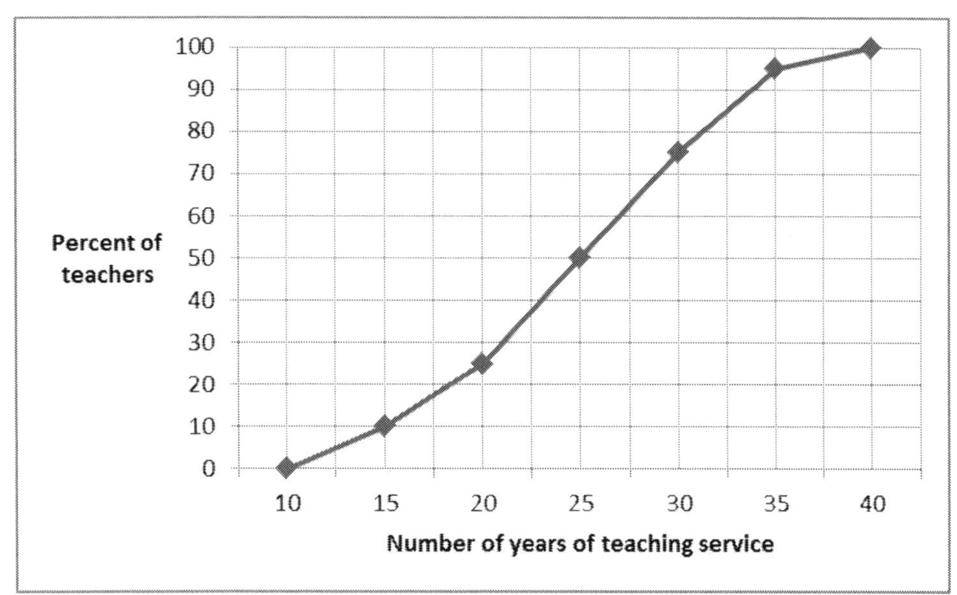

(1) The median for the number of years of teaching service is 25 ☐

(2) 75% of teachers serve more than 30 years ☐

(3) The interquartile range for the teaching service is 10 years ☐

Question 12

Five classes raise money for a charity. The amount raised per class as well as the number of pupils in each class is shown in the table below:

Point and click on the class where the amount raised per pupil was the highest, working out your answers to the nearest penny. (In this case, in the last column, put an 'X' in the appropriate row that corresponds to the highest amount raised per pupil)

Class	Number of Pupils	Amount of money raised in £	Put an X in the row that corresponds to the highest amount raised per pupil
A	27	37.50	
B	22	32.75	
C	28	40.20	
D	21	22.50	
E	24	24.80	

Question 13

A pupil scores 15 out of 60 marks in Test A and 45 out of 80 marks in Test B. The tests are weighted as follows. 60% is given for Test A and 40% for Test B. Given these weightings what is the overall percentage score the pupil gains?

Answer: 

Question 14

The Deputy Head created the following table showing the number of pupils in each year group who had music lessons.

What is the percentage of pupils in all the year groups combined that have music lessons? Give your answer rounded to a whole number.

Year Group	Number of pupils	Number of pupils who have music lessons
7	92	10
8	101	18
9	105	14
10	96	13
11	102	11

Answer: ☐ %

Question 15

In a certain school the number of pupils entered for GCSE English was 320. The number of boys in this group was 240. What was the proportion of girls to boys that entered GCSE English in this school? Choose the correct value from the four options given below

.

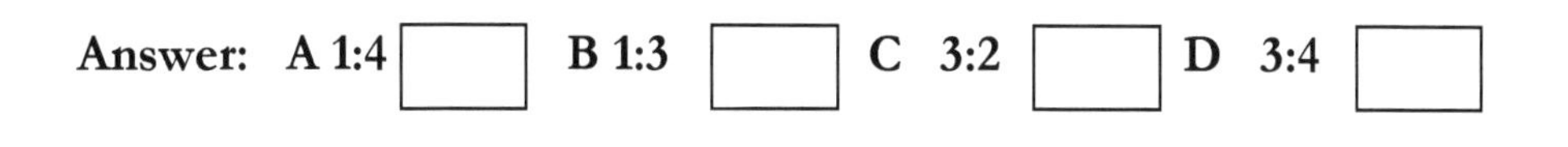

Answer: A 1:4 ☐ B 1:3 ☐ C 3:2 ☐ D 3:4 ☐

Question 16

In one school 140 pupils took maths GCSE exams. Both the percentage of pupils as well as the GCSE Grades obtained in maths is shown in the pie chart below. Indicate all true statements

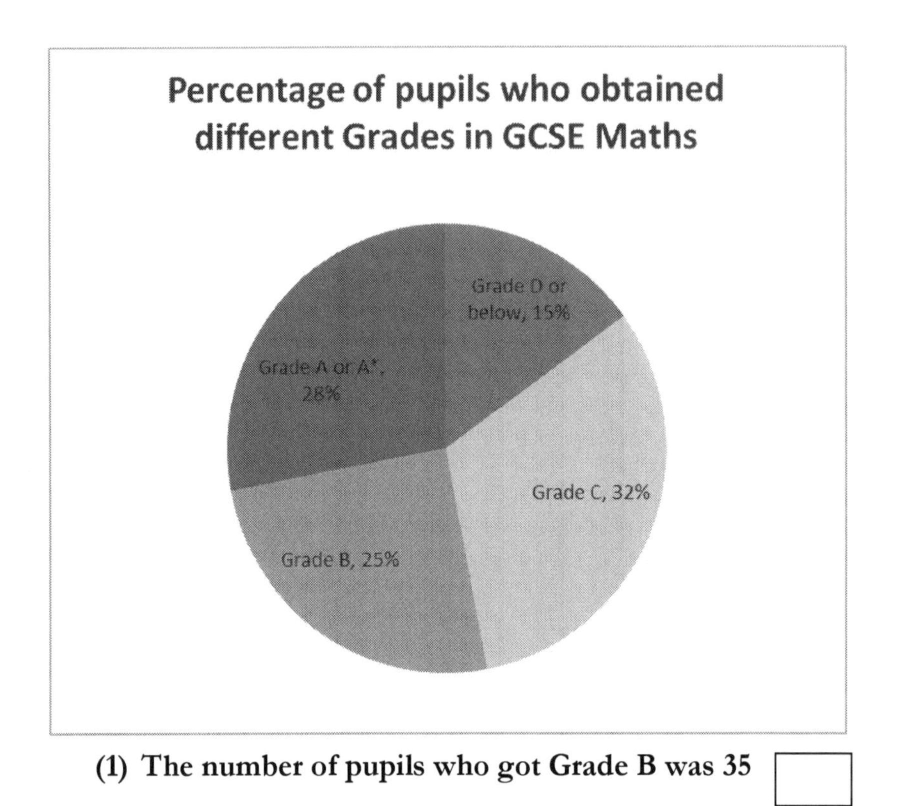

(1) The number of pupils who got Grade B was 35 ☐

(2) The percentage of pupils who got Grade C or above was 85% ☐

(3) The number of pupils who obtained Grade A or A* was 28 ☐

Answers to Mock On screen Test 2

Question 1

Answer £174.10

Method: The total number of people that go on this trip = 22 pupils + 4 teachers = 26 people. The cost of tickets for 26 people = 26×£4.85 = £126.10. Since the cost of return travel by coach = £48, the total cost of the trip = £126.10 +£48 = £ 174.10

Question 2

Answers

(1) False – since school B also outperforms school A in 2007

(2) True

(3) False – in 2006 school B performs better than school A

Question 3

Answer: 21:27 hours

Method: Total time taken for appointments = 18×14 =252 minutes

Total time including breaks = 252 + 30 = 282 minutes = 240 + 42 = 4 hours 42 minutes

So the parents evening finished at 16:45 + 4hhrs and 42mins = 21:27

Question 4

Answers

(1) False – The range in 2010 = 85 – 15 = 70, range in 2011 = 75 – 20 = 55, the range in 2010 is bigger

(2) True

(3) False – IQR in 2011 = 65 – 30 = 35, IQR in 2010 = 60 -20 =40, so the IQR in 2011 is smaller than 2010

Question 5

Answer £20.65

Method: Pupil gets 80X1.5 = 120 CHF, he spends 88 CHF, so has 120 – 88 = 32 CHF left

32 CHF = 32 ÷ 1.55 = £20.65

Question 6

Answer: 14:15 hours

Method: 25 miles per hour = $\frac{8}{5} \times 25$ = 40 kilometres per hour

Time taken to go to Versailles = Distance ÷ Speed = 30÷40 =$\frac{3}{4}$ hrs

= 45 minutes

Since she wants to be in Versailles at 15:00 hrs, she needs to leave Paris at 14:15

Question 7

Answer: 340 pupils

The percentage of boys that achieve at least a level 5 in KS3 from the 850 pupils in English is 40%

40% of 850 pupils = $\frac{40}{100} \times 850$ = 340 pupils

Question 8

Answers:

(1) True

(2) False – The number of pupils who get 50 marks or more in the test is 6 not 7

(3) True

Question 9

Answer: School B

The school B is the only school that consistently shows a trend improvement, in this case by 0.3 per year (from 26.6 to 26.9 = 0.3 increase, from 26.9 to 27.2 = 0.3 increase, etc)

Question 10

Answer: 12 pupils

Method: Since there are 54 pupils altogether and 24 are girls, this means 30 are boys

Pupils that take music lessons are: boys = $\frac{1}{5} \times 30 = \frac{30}{5} = 6$ and

girls= $\frac{1}{4} \times 24 = \frac{24}{4} = 6$

Hence total number of pupils that take music lesson = 12

Question 11

Answers:

(1) True

(2) False – only 25% of teachers have more than 30 years of service

(3) True – IQR = Upper quartile – Lower Quartile = 30 – 20 =10

Question 12

Answer: Class B

Method: work out the amount raised per pupil for each class by dividing the total for each class by the number of pupils. Class A = £1.39, Class B = £1.49, Class C = £1.44, Class D = £1.07 and class E = £1.03. So class B corresponds to the highest amount raised per pupil.

Question 13

Answer: 37.5%

Method: In Test A, the pupil scores 15 out of 60 = $\frac{15}{60}$ × 100 = $\frac{1}{4}$ ×

100 = 25%

Similarly in Test B, the pupil scores 45 out of 80 = $\frac{45}{80}$ × 100 =

56.25%

Now applying the appropriate weightings we have the overall percentage = 60% of 25% plus 40% of 56.25%. This means 0.6 × 25 + 0.4X56.25 = 15% + 22.5% = 37.5%

Question 14

Answer = 13%

Method: The total number of pupils in all the year groups combined = 496. The total number of pupils who take music lessons = 66. So the percentage of pupils who take music lessons =$\frac{66}{496}$ × 100 = 13% (rounded to a whole number)

Question 15

Answer: 1:3

Method: There are 320 pupils who entered GCSE English of whom 240 were boys. This means 80 of this group were girls. The ratio of girls to boys = 80:240 which simplifies to 8:24 (divide both 80 and 240 by 10), finally 8:24 simplifies to 1: 3 (divide both 8 and 24 by 8)

Question 16

Answer:

(1) True – since 25% of 140 = 35

(2) True – Since 32% + 25% + 28% = 85%

(3) False – since 28% of 140 = 39 and not 28

Some basic and interesting reminders:

I am sure you know what Even and Odd numbers are, but there is something fascinating about odd numbers which may interest you. See below:

Even Numbers: All numbers that have 2 as a factor are even numbers.

Examples are: 2, 4, 6, 8, 10, 12, 14, 16

So, 168 is an even number as it can be divided exactly by 2.

Odd Numbers: Are all numbers that do not have 2 as a factor.

Examples: 1, 3, 5, 7, 9, 11, 13, 15, 17

So for example 81 is an odd number, as 2 is not a factor of 81.

Notice something interesting about odd numbers. The cumulative sum of consecutive odd numbers generates square numbers.

The first number is $1 = 1 \times 1 = 1^2$

The sum of the first two numbers is $1 + 3 = 4 = 2 \times 2 = 2^2$

The sum of the first three numbers is $1 + 3 + 5 = 9 = 3 \times 3 = 3^2$

The sum of the first four numbers is $1 + 3 + 5 + 7 = 16 = 4 \times 4 = 4^2$ and so on.

Multiples: These are simply numbers in the multiplication tables.

For example the multiples of 6 are 6, 12, 18, 24, 30,.....

Factors: A factor is a number that divides exactly into another number for example, the number 2 in the case of even numbers.

3 is a factor of 9, as 3 goes exactly into 9.

15, has two factors other than 15 and 1. The two factors are 5 and 3, since both these numbers go exactly into 15.

Example: Find all the factors of 21. The factors are: 1, 3, 7 and 21

Prime Numbers: A prime number is a number that has two factors, the number itself and 1. Examples of prime numbers include 2, 3, 5, 7, 11, 13 and 17. So for example 23 is also a prime number since it has no other factor besides itself and 1.

Another example: Find the 3rd prime number after 11. The first prime after 11 is 13, the second prime is 17, and the third prime after 11 is 19.

Reciprocals:

The reciprocal of a number is 1 divided by the number. For example the reciprocal of 5 is $\frac{1}{5}$, the reciprocal of 13 is $\frac{1}{13}$ and so on.

Divisibilty by 9:

A number is divisible by 9 if its reduced digit sum is 9. Example: the number 18 can be reduced to 1+8 =9, so 18 is divisible by 9. 567 can be reduced to 5 +6 +7 =18 and 18 =1 + 8 =9. So, 567 is divisible by 9. 59049 can be reduced to 5 +9 +0 + 4 +9 = 27 and 27 = =2 +7 =9. Hence, the number, 59049 can be divided by 9.

Useful Website Addresses

Practice Material

(1) Online assessments:
http://www.education.gov.uk/QTS/Numeracy/assessment_engine.html

(2) Practice paper 2:
http://media.education.gov.uk/assets/files/pdf/n/numeracy%20practice%20paper%202.pdf

(3) Practice paper 3:

http://media.education.gov.uk/assets/files/pdf/n/numeracy%20practice%20paper%203.pdf

Current details of test contents:

http://www.education.gov.uk/schools/careers/traininganddevelopment/professional/b00211213/numeracy/content

Information on skills test registration:

http://www.education.gov.uk/schools/careers/traininganddevelopment/professional/b00211200/registration

Other Books that may be useful: **Speed Mathematics for Primary school teachers** – suitable for primary school teachers and parents who want to teach kids (7-11 year olds) to do speed mathematics

See next page for discount for online tutorials.

Printed in Great Britain
by Amazon